X.media.publishing

For further volumes:
http://www.springer.com/series/5175

M.R.C. van Dongen

LaTeX and Friends

 Springer

Dr M.R.C. van Dongen
Computer Science Department
University College Cork
Cork
Ireland

ISSN 1612-1449
ISBN 978-3-642-23815-4   ISBN 978-3-642-23816-1
DOI 10.1007/978-3-642-23816-1
Springer Heidelberg Dordrecht London New York

Library of Congress Control Number: 2011945089

Printed on acid-free paper

Springer is part of Springer Science+Business Media (www.springer.com)

Foreword

NEARLY TWENTY YEARS after the first ideas for LaTeX2ε emerged, the use of LaTeX to produce high-quality technical documents shows no sign of waning. Indeed, over the past 5 or so years there has been if anything an *upturn* in interest in using LaTeX. Better editors, faster computers and the range of powerful LaTeX packages have all contributed to this increased uptake.

For the new user, this vibrancy can appear intimidating. The range of packages available for use with LaTeX is vast, and it is not always obvious which is the 'best of breed.' What new users need therefore is a guide not just to the basics of the LaTeX approach, but also help in navigating this ecosystem so that they can produce the documents they need as rapidly as possible.

Creating well-designed documents is about more than the technical detail of any typesetting system, and so as well as learning LaTeX it is also necessary to understand the wider ideas of good writing and good design if one is to create truly 'beautiful' material.

In *LaTeX and Friends*, Marc van Dongen provides an integrated solution to these inter-related requirements. Treating the presentation of beautiful documents as the key aim of the reader, it offers advice on good practice (both in LaTeX terms and beyond) in the relevant context for the beginner. It also avoids the problem seen in many texts, which fall short in supporting the transition from beginner to advanced user. Thus while new LaTeX users will find the information they need here, so will more established users, making this not only a beginners' guide but also a reference manual for day-to-day LaTeX users.

Joseph Wright

Contents

List of Figures

List of Tables

Preface

THIS BOOK PROVIDES students with an introduction to technical writing and computer presentations with LaTeX, which is the de-facto standard in computer science and mathematics. The book may also be used as a reference for seasoned LaTeX users.

The book offers techniques for writing large and complex documents, preparing computer presentations, and creating complex graphics in an integrated manner. The book's website, which may be found at http://csweb.ucc.ie/~dongen/LAF, has three separate chapters explaining how to use a widely used LaTeX distribution on Windows, on Unix, and on the Mac. These chapters also provide an introduction to some selected integrated development environments (IDES).

I have tried to minimise the number of classes and style files the reader has to know. This is one of the main reasons why I decided to use the amsmath package for the presentation of mathematics, and decided to use tikz, pgfplots, and beamer for the creation of diagrams, data plots, and computer presentations. Another advantage of this approach is that it simplifies the process of creating a viewable/printable output file because everything should work with pdflatex, which is a program that turns LaTeX into pdf.

The book avoids the use of what is known in the LaTeX community as "verbatim" commands and environments. except when it comes to including, well, verbatim program listings. The main reason for this decision is that verbatim commands in the hands of beginners often lead to errors that are difficult to find and are not always so easy to resolve. By no means should the decision to omit verbatim commands be a limitation; this book was written without verbatim commands, so why should you need them when you're writing a thesis or dissertation?

M.R.C. van Dongen
Cork
2011

Book Outline

THIS BOOK has seven parts, some of which are more technical than others. The following is a short outline.

The first two parts are called *Basics* and *Basic Typesetting*. These parts introduce the reader to the basic LaTeX commands for typesetting and cross-referencing. They also explain how to create one or several bibliographies and one or several indexes or glossaries.

The next part is *Tables, Diagrams, and Data Plots*, which is about presenting data in tables, diagrams with the `tikz` package, and data plots with the `pgfplots` package.

Mathematics and Algorithms is the next part. It explains how to typeset mathematics, how to typeset algorithms in pseudo-code, and how to present program listings.

This is followed by *Automation*, which explains how to implement user-defined commands, how to implement option parsing, and how to implement conditional branching. Some readers may wish to skip this part because it is more technical than the other parts.

Miscellany is the next part. It is a collection of optional chapters, some of which are of a more technical nature than others. The first, relatively easy, chapter explains how to create computer presentations with the `beamer` package. It continues with two more technical chapters that explain how to implement user-defined classes and packages and how to use OpenType fonts.

The last part is *References and Bibliography*, which is a collection of indexes, a list of acronyms, a bibliography, and a short typographic jargon reference. Readers not familiar with notions such as characters, glyphs, ligatures, serifs, kerning, fonts, typefaces, points, point size and leading, ems, and ens, are invited to start with the jargon reference before reading the rest of the book.

Overall, the chapters are well balanced but the chapters about typesetting mathematics and presenting diagrams with `tikz` are a bit longer and more detailed. This is why it was decided to split the presentation on typesetting mathematics into two separate chapters. The first of these chapters should be sufficient for most readers. The chapter about presenting diagrams with `tikz` was not split because it was felt that most readers who are interested in some of this chapter would also be interested in the rest.

Acknowledgements

THIS BOOK would not have been possible without the help of many. First of all, I should like to thank Don Knuth for writing TEX and Leslie Lamport for writing LATEX—without them the landscape of computer-based typesetting would have been dominated by Bill. I should like to thank Eddie Kohler for writing otftotfm and for his help. I am grateful to Till Tantau and colleagues for writing the beautiful tikz package and the beamer class. Both of them are stars in terms of functionality, productivity, and documentation. Thanks to David Farley and Dario Taraborelli for letting me include the pictures in Figures 4.2 and 16.1. Many thanks to Billy Foley and the University College Cork Art Collection for letting me include the pictures at the back of the part titlepages. I should like to thank Frank Böhme, George Boyle, Tom Carroll, Hans Hagen, Taco Hoekwater, Finbarr Holland, Rik Kabel, Mico Loretan, Ben McKay, Luca Mercriadri, Oliver Nash, Oleg Paraschenko, Jason Quinlan, Lisa Swenson, and Uwe Ziegenhagen for useful comments on early drafts. I should also like to thank Paul Blaga, Robin Fairbairns, Peter Flynn, Francisco A. F. Reinaldo, and Boris Veytsman for reviewing the book. Special thanks to Joseph Wright who was so kind to proofread the entire book and to write the foreword. His critical eye spotted many known and unknown errors. Many thanks to Mr Engesser, Ms Glaunsinger, and Ms Fisher at Springer for providing the opportunity to publish this book and for helping me bring this project to a successful end. Finally, I should like to thank all those who have worked on LATEX and friends, all those who have supported LATEX and friends, and all who have answered all my LATEX and METAPOST questions over the last two decades or so. The following are but a few: André Heck, Barbara Beeton, Cristian Feuersänger, Dan Luecking, David Carlisle, David Kastrup, Denis Roegel, Donald Arseneau, D. P. Story, Frank Mittelbach, Frank van Raalte, Hans Hagen, Heiko Oberdiek, Jim Hefferon, John Hobby, Jonathan Fine, Jonathan Kew, Karl Berry, Kees van der Laan, Keith Reckdahl, Kjell Magne Fauske, Mark Wibrow, Nelson Beebe, Peter Wilson, Philipp Lehman, Rainer Schöpf, Ross Moore, Scot Pakin, Sebastian Rahtz, Stephan Hugel, Taco Hoekwater, Thomas Esser, Ulrike Fisher, Victor Eijkhout, Vincent Zoonekynd, Will Robertson, and all the many, many others. Without them the TEX community would have been much worse off.

PART I

Basics

Untitled Landscape, oil on paper (1993), 64 × 90 cm
Work included courtesy of Billy Foley and University College Cork Art Collection
© Billy Foley (www.billyfoley.com) and University College Cork Art Collection

CHAPTER 1
Introduction to LaTeX

THIS CHAPTER is an introduction to LaTeX and friends but it is *not* about typesetting fancy things. Typesetting fancy things is covered in subsequent chapters. The main purpose of this chapter is to provide an understanding of the *basic* mechanisms of LaTeX, using plain text as a vehicle. Having read this chapter you should know how to:

○ Write a simple LaTeX input document based on the `article` *class*.
○ Turn the input document into `pdf` with the `pdflatex` program.
○ Define *labels* and use them to create consistent cross-references to chapters and sections. This basic cross-referencing mechanism also works for tables, figures, and so on.
○ Create a fault-free table of contents with the `\tableofcontents` command. Creating a list of tables and a list of figures works in a similar way.
○ Cite the literature with the aid of the `\cite` command.
○ Generate one or several bibliographies from your citations with the `bibtex` program.
○ Change the appearance of the bibliographies by choosing the proper bibliography style.
○ Manage the structure and writing of your document by exploiting the `\include` command.
○ Control the visual presentation of your article by selecting the right `article` class options.
○ Much, much, more.

Intermezzo. LaTeX gives you output documents that look great and have consistent cross-references and citations. Much of your output document is created automatically and much is done behind the scenes. This gives you extra time to think about the ideas you want to present and how to communicate these ideas in an effective way. One route leading to effective communication is planning: the order and the purpose of the writing determines how it is received by your target audience. LaTeX's markup helps you concretise the purpose of your writing and present it in a consistent manner. As a matter of fact, LaTeX *forces* you to think about the purpose of your writing and this improves the effectiveness of the presentation of your ideas. All that's left for you is to determine the order of presentation and provide some extra markup. To determine the order of your presentation and to write your document you can treat

LaTeX as a programming language. This means that you can use software engineering techniques such as top-down design and stepwise refinement. These techniques may also help when you haven't completely figured out what it is you want to write.

This chapter requires that you do a few things at the Operating System (OS) level. Throughout is assumed that you use the unix operating system. The online chapters, which may be found at http://csweb.ucc.ie/~dongen/LAF, explain how to carry out these OS-related tasks on Windows and on the Mac.

1.1 Pros and Cons

Before we start, it is good to look at arguments in favour of LaTeX and arguments against it. Some of these arguments are based on comments from Taraborelli [2010]. The following are some common and less common arguments against LaTeX.

○ LaTeX is difficult. It may take one to several months to learn. True, learning LaTeX does take a while. However, it will save you time in the long run, even if you're writing a minor thesis.
○ LaTeX is not a What You See is What You Get (WYSIWYG) wordprocessor. Correct, but there are many LaTeX Integrated Development Environments (IDEs) and some IDEs such as eclipse have LaTeX plugins.
○ There is little support for physical markup. Yes, but for most papers, notes, and theses in computer science, mathematics, and other technical and non-technical fields, there are existing packages that you can use without having to fiddle with the way things look. However, if you really need to tweak the output then you may have to put in extra time, which may slow down the writing. Then again, you should be able to reuse this effort for other projects.
○ Using non-standard fonts is difficult. This used to be true. However, with the arrival of the fontspec package and X3TEX using non-standard fonts is easy. Furthermore, it is more than likely that for most day-to-day work you wouldn't *want* any non-standard fonts.
○ It takes some practice to let text flow around pictures. That's a tricky one. Usually LaTeX determines the positions of your figures. As a consequence the figures may not end up where you want. Sometimes the text in the vicinity of such figures doesn't look nice: the text doesn't flow. You can improve the text flow by rearranging a few words in adjacent paragraphs but this *does* take some practice.
○ LaTeX doesn't provide spell checking. That is true, but most modern IDEs have spell checkers. Furthermore, there are command line tools such as ispell that can spell check LaTeX input.
○ There are too many LaTeX packages, which makes it difficult to find the right package. Agreed, but most LaTeX documents only require a few packages that are easy to find. Moreover, asking a question in the mailing list comp.text.tex (also in http://groups.google.com/group/comp.text.tex/topics) usually results in some quick pointers.

○ LaTeX encourages structured writing and the separation of style from content, which is not how everybody works. Well, it seems times they are-a-changin' because more and more new (new?) communities have started using LaTeX [Burt 2005; Thomson 2008a; Thomson 2008b; Buchsbaum, and Reinaldo 2007; Garcia, and Buchsbaum 2010; Breitenbucher 2005; Senthil 2007; Dearborn 2006; *LaTeX: A Guide for Philosophers*; Veytsman, and Akhmadeeva 2006]. Some communities have organised and have their own websites [Schneider 2011; *LaTeX for Logicians*; *PhilTeX Forums* A Place for Philosophers to Learn about LaTeX; Ochsenmeier 2011; Dashboard 2008; Arnold 2010].

The following are arguments in favour of LaTeX. If you're not familiar with technical phrases such as kerning, small caps, ligatures, and so on then don't worry, they will be explained further on. Also remember that typographic notions are explained at the end of the book.

○ LaTeX provides state-of-the art typesetting, including kerning, real small caps, common and non-common ligatures, glyph variants, ... It also does a very good job at automated hyphenation.
○ Many conferences and publishers accept LaTeX. In addition they provide classes and packages that guarantee documents conforming to the required formatting guidelines.
○ LaTeX is a Turing-complete programming language. This gives you almost complete control. For example, you can decide which things should be typeset and how this should be done.
○ With LaTeX you can prepare several documents from the same source file. Not only can you control which text should be used in which document but also how it should appear.
○ LaTeX is highly configurable. To change the appearance of your document you choose the proper document class, class options, packages, and package options. The proper use of commands guarantees a consistent appearance and gives you ultimate control.
○ You can translate LaTeX to `html/ps/pdf/DocBook`....
○ LaTeX automatically numbers your chapters, sections, figures, and so on. In addition it provides cross-referencing support.
○ LaTeX has excellent bibliography support. It supports consistent citations and an automatically generated bibliography with a consistent look and feel. The style of citations and the organisation of the bibliography is configurable.
○ There is some support for WYSIWYG document preparation: `lyx` (`http://www.lyx.org/`), TeXmacs (`http://www.texmacs.org/`), ... Furthermore, some editors and IDEs provide support for LaTeX, e.g., `vim`, `emacs`, `eclipse`, ...
○ LaTeX is *very* stable, free, and available on many platforms.
○ There is a very large, active, and helpful TeX/LaTeX user-base.
○ LaTeX has comments.
○ Most importantly: LaTeX is fun!

1.2 Basics

LaTeX [Lamport 1994] was written by Leslie Lamport as an extension of Donald Knuth's TeX program [Knuth 1990]. It consists of a Turing-complete procedural markup language and a typesetting processor. The combination of the two lets you control both the visual presentation *as well as* the content of your documents. The following three steps explain how you use LaTeX.

1. You write your document in a LaTeX (`.tex`) input (source) file.
2. You run the `latex` or `pdflatex` program on your input file. This turns the input file into a device independent (`.dvi`) file or a portable document format (`.pdf`) file. If there are errors in your source file then you should fix them before you can continue to Step 3.
3. You view the `dvi` or `pdf` file on your computer. A `dvi` file cannot be printed directly. You have to convert it to postscript or `pdf` and then print the result of the conversion. The advantage of `pdf` is that it can be printed straight away.

1.2.1 *The TeX Processors*

Roughly speaking LaTeX is built on top of TeX. This adds extra functionality to TeX and makes writing your document much easier. With LaTeX being built on top of TeX, the result is a TeX program. You may get a good understanding of LaTeX by studying TeX's four *processors*, which are basically run in a "pipeline" [Knuth 1990; Eijkhout 2007; Abrahams, Hargreaves, and Berry 2003]. The following are the main functions of TeX's processors.

Input Processor Turns the source file into a token stream. The resulting token stream is sent to the Expansion Processor.

Expansion Processor Turns the token stream into token stream. Expandable tokens are repeatedly expanded until there are no more left. The expansion applies to commands, conditionals, and some primitive commands. The resulting output is sent to the Execution Processor.

Execution Processor Executes executable control sequences. These actions may affect the state. This applies, for example, to assignments and command definitions. The Execution Processor also constructs horizontal, vertical, and mathematical lists. The final output is sent to the Visual Processor.

Visual Processor Creates the `dvi` or `pdf` file. This is the final stage. It turns horizontal lists into paragraphs, breaks vertical lists into pages, and turns mathematical lists into formulae.

1.2.2 *From tex to dvi and Friends*

Now that you know a bit about how LaTeX works, it's time to study the programs you need to turn your input files into readable output. You may ignore this section if you use an IDE because your IDE will do all the necessary things to create your output file, without the need for user intervention at the command-line level.

In its simplest form the `latex` program turns your LATEX input file into the device independent (`dvi`) file, which you can view and turn into other output formats, including `pdf`. Before going into the details about the LATEX syntax, let's see how you turn an existing LATEX source file into a `dvi` file. To this end, let's assume you have an error-free LATEX source file that is called ⟨base name⟩`.tex`. The following command turns your source file into an output file called ⟨base name⟩`.dvi`.

```
$ latex ⟨base name⟩.tex                                    Unix Session
```

With `latex` you may omit the `tex` extension.

```
$ latex ⟨base name⟩                                        Unix Session
```

The resulting `dvi` output can be viewed with the `xdvi` program.

```
$ xdvi ⟨base name⟩.dvi &                                   Unix Session
```

Now that you have the `dvi` version of your LATEX program, you may convert it to other formats. The following converts ⟨base name⟩`.dvi` to postscript (⟨base name⟩`.ps`).

```
$ dvips -o ⟨base name⟩.ps ⟨base name⟩.dvi                  Unix Session
```

The following converts ⟨base name⟩`.dvi` to portable document format (`pdf`).

```
$ dvipdf ⟨base name⟩.dvi                                   Unix Session
```

However, by far the easiest to generate `pdf` is using the `pdflatex` program. As with `latex`, `pdflatex` does not need the `tex` extension.

```
$ pdflatex ⟨base name⟩.tex                                 Unix Session
```

Intermezzo. If you're writing a book, a thesis, or an article then generating `dvi` and viewing it with `xdvi` is by far the quickest. However, there may be problems with graphics, which may not always be rendered properly. I find it convenient to (1) run the `xdvi` program in the background (with the & operator), (2) put the `xdvi` window on top of the window that edits the LATEX program, and (3) edit the program with `vim`. You can execute shell commands from within `vim` by going to command mode and issuing the command. For example, ⟨ESC⟩:!⟨command⟩⟨RETURN⟩ executes the command ⟨command⟩ and ⟨ESC⟩:!!⟨RETURN⟩ executes the most recently executed command.[1] This lets you run `latex` on your input document from within your editor. Most Linux Graphical User Interfaces (GUIs) let you cycle from window to window if you type a magic spell: in KDE it is ⟨ALT⟩⟨TAB⟩. Typing this spell lets me quickly cycle from my editing session to the viewing sessions and back. Using this mechanism keeps my hands on the keyboard and saves time, wrists, and elbows.

---

[1] The `emacs` program should let you to do similar things.

1.2.3 *The Name of the Game*

Just like C, lisp, pascal, java, and other programming languages, LaTeX may be viewed as a program or as a language. When referring to the language this book usually uses LaTeX and when referring to the program it usually writes latex. However, when writing latex the book actually means pdflatex because this is by far the easiest way to create viewable and printable output. Finally, when this book writes 'LaTeX program' it usually means 'LaTeX source file.'

1.2.4 *Staying in Sync*

Sometimes the latex program needs more than a single run to produce its final output. The following explains what happens when you and latex are no longer in sync.

To create a perfect output file and have consistent cross-references and citations, latex also writes information to and reads information from *auxiliary* files. Auxiliary files contain information about page numbers of chapters, sections, tables, figure, and so on. Some auxiliary files are generated by latex itself (e.g., aux files). Others are generated by external programs such as bibtex, which is a program that generates information for the bibliography. When an auxiliary file changes then LaTeX may be out of sync. You should rerun latex when this happens. Normally, latex outputs a warning when it suspects this is required:

```
                                                    Unix Session
$ latex document.tex
... LaTeX Warning: Label(s) may have changed. ...
Rerun to get cross-references right.
$
```

1.2.5 *Writing a LaTeX Input Document*

LaTeX is a markup language and document preparation system. It forces you to focus on the content and *not* on the presentation. In a LaTeX program you write the content of your document, you use commands to provide markup and automate tasks, and you import libraries. The following explains this in further detail.

content The content of your document is determined in terms of text and logical markup. LaTeX forces you to focus on the logical structure of your document. You provide this structure as markup in terms of familiar notions such as the author of the document, the title of a section, the body and the caption of a figure, the start and end of a list, the items in the list, a mathematical formula, a theorem, a proof, ...

commands The main purpose of commands is to provide markup. For example, to specify the author of the document you write \author{⟨author name⟩}. The real strength of LaTeX is that it also is a Turing-complete programming language, which lets you define your own commands. These commands let you do real programming and give you ultimate

```
\documentclass[a4paper,11pt]{article}

%␣Use␣the␣mathptmx␣package.
\usepackage{mathptmx}

\author{A.\,U.␣Thor}
\title{Introduction␣to␣\LaTeX}
\date{\today}

\begin{document}␣%␣Here␣we␣go.
␣␣\maketitle
␣␣\section{Introduction}
␣␣␣␣␣The␣start.
␣␣\section{Conclusion}
␣␣␣␣␣The␣end.
\end{document}
```

Figure 1.1
Typical LATEX program

control over the content and the final visual presentation. You can reuse your commands by putting them in a library.

libraries There are many existing document classes and packages (style files). Class files define rules that determine the appearance of the output document. They also provide the required markup commands. Packages are best viewed as libraries. They provide useful commands that automate many tedious tasks. However, some packages may affect the appearance of the output document.

Throughout this book, LATEX input is typeset in a style that is reminiscent of the layout of a computer programming language input file. The style is very generous when it comes to inserting redundant space characters. Whilst not strictly necessary, this input layout has several advantages:

show structure Carefully formatting the input shows the structure of your LATEX source files. This makes it easier to locate the start and end of sentences and higher-level building blocks such as *environments*. (Environments are explained further on.)

mimic output By formatting the input you can mimic the output. For example when you design a table with rows and columns you can align the columns in the input. This makes it easier to design the output.

find errors This is related to the previous item. Formatting may help you find the cause of errors more quickly. For example, you can reduce the number of candidate error locations by commenting out entire lines. This is much easier than commenting out parts of lines, which usually requires many more editing operations. Especially if your editor supports "multiple undo/redo operations" this makes locating the cause of errors very easy.

Figure 1.1 depicts a typical example of a LATEX input program. For this example all spaces in the input are made explicit by typesetting them with the symbol '␣,' which represents a single space. The symbol ␣ is called *visible space*. In case you're wondering, the command \textvisiblespace typesets the visible space.

The remainder of this section studies the example program in more detail. Spaces are no longer made explicit.

The third line in the input program is a comment. A comment starts with a percentage sign (%) and lasts until the end of the current line. Comments, as is demonstrated in the input program, may also start in the middle of a line.

The following command tells LATEX that your document should be typeset using the rules determined by the article document class.

```
\documentclass[a4paper,11pt]{article}
```
LATEX Usage

You can only have one document class per LATEX source file. The \documentclass command determines the document class. The command takes one *required* argument, which may be a single character or a sequence of characters inside braces (curly brackets). The argument is the name of the document class. In our example the required argument is article so the document class is article. You usually use the \documentclass command on the first line of your LATEX input file.

In our example, the \documentclass command also takes an *optional* argument. An optional argument is passed inside the square brackets immediately after the command (this is standard). Optional arguments are called *optional* because they may be omitted. If you omit them then you should omit the square brackets. In our example the 'a4paper,11pt' are options of the \documentclass command. The \documentclass command passes these options to the article class. This sets the default page size to A4 with wide margins and sets the font size to 11 point.

The following command includes a package called mathptmx.

```
\usepackage{mathptmx}
```
LATEX Usage

The mathptmx package sets the default font to *Times Roman*. This is a very compact font, which may save you precious pages in the final document. Using the font is especially useful when you're fighting against page limits.

Packages may also take options. This works just as with document classes. You pass the options to the package by including them in square brackets after the \usepackage command.

The following three commands, which are best used in the preamble, are logical markup commands. These commands do not produce any output but they define the author, title, and date of our article.

```
\author{A.\,U. Thor}
\title{Introduction to \LaTeX}
\date{\today}
```
LATEX Usage

The command \LaTeX in the argument of the \title command is for typesetting LATEX. The command \, inserts a litte space. The amount of space is called *thin space*. Bringhurst [2008] recommends that you add a thin space (or no space) within strings of initials.

The title of your document is typeset with the `\maketitle` command. Usually, you put this command at the start of the `document` *environment*, which is the text between `\begin{document}` and `\end{document}`. In the argument of the `\author` command the author names should be separated with the `\and` command:

```
\author{Donald E. Knuth \and Peter B. Bendix}
```
LATEX Usage

You acknowledge friends, colleagues, and funding institutions by including a `\thanks` command as part of the argument of the `\author` command. This produces a footnote consisting of the argument of the `\thanks` command.

```
\author{Sinead\thanks{You're a luvely audience.}}
```
LATEX Usage

If you wish to build your own titlepage, then you may do this with the `titlepage` environment. This environment gives you complete control *and* responsibility. The `\titlepage` command and the `titlepage` environment may only be used inside the `document` environment (between the `\begin{document}` and `\end{document}`).

For the `article` class, as well as for most other LATEX classes, you write the main text of the document in the `document` *environment*. This environment starts with `\begin{document}` and ends with `\end{document}`. We say that text is "in" the `document` environment if it is between the `\begin{document}` and `\end{document}`. The text before `\begin{document}` is called the *preamble* of the document. Sometimes we call the text in the `document` environment the *body* of the document environment.

Definitions and configurations should be provided in the preamble. The text in the `document` environment defines the content. In the body of your document you may use the commands that are defined in the preamble. (More generally, you may define commands almost anywhere. You may use them as soon as they're defined.)

The body of the `document` environment of the example in Figure 1.1 defines a rather empty document consisting of a title, two sections, and two sentences. The title is generated by the `\maketitle` command. The sections are defined with the `\section` command. Each section contains one sentence. The text 'The start.' is in the first section. The text 'The end.' is in the last.

1.2.6 *The Abstract*

Many documents have an *abstract,* which is a short piece of text describing what is in the document. Typically, the abstract consists of a few lines and a few hundred words. You specify the abstract as follows.

```
\begin{abstract}
    This document is an introduction to \LaTeX. ...
\end{abstract}
```
LATEX Usage

Figure 1.2

Defining comments

```
This is the␣␣first sentence
of the first paragraph.
The second sentence of this
paragraph ends in the word
'elephant.'

This is the first sentence
of the second pa%comment
ragraph.
The second sentence of this
paragraph
ends in the word '%eleph
ant.'
```

This is the first sentence of the first paragraph. The second sentence of this paragraph ends in the word 'elephant.'

This is the first sentence of the second paragraph. The second sentence of this paragraph ends in the word 'ant.'

In an article the abstract is typically positioned immediately after the \maketitle command. Abstracts in books are usually found on a page of their own.

Some classes may provide an \abstract command that defines the abstract. These classes may require that you use the \abstract command in the document preamble.

1.2.7 Spaces, Comments, and Paragraphs

The paragraph is one of the most important basic building blocks of your document. The paragraph formation rules depend on how latex treats spaces, empty lines, and comments. Roughly, the rules are as follows.[2] In its default mode, latex treats a sequence of one or more spaces as a single space. The end of the line is the same as a space. However:

- An empty line acts as an end-of-paragraph specifier.
- A percentage character (%) starts a comment that ends at the end of the line.
- Spaces at the start of a line following a comment are ignored.

If you understand the example in Figure 1.2 then you probably understand these rules. In this example, the input is on the left and the resulting output on the right. This convention is used throughout this book, except for Chapter 5, which presents pictures on the left and LaTeX input on the right.

1.3 Document Hierarchy

The coarse-level logical structure of your document is formed by the parts in the document, chapters in parts, sections in chapters, subsections in sections, subsubsection in subsections, paragraphs, and so on. This defines the *document hierarchy*. Following [Lamport 1994], we shall refer to the members of the hierarchy as *sectional units*.

---

[2]Here it is assumed that the text does not contain any commands.

Intermezzo. The sectional units are crucial for presenting *effectively*. For example, you break down the presentation of a thesis by giving it chapters. The chapters should be ordered to ease the flow of reading. The titles of the chapters are also important. Ideally chapter titles should be short, but most importantly each chapter title should describe what's in its chapter. To the reader a chapter title is a great help because it prepares them for what's in the chapter that they're about to read. A good chapter title is like an ultimate summary of the chapter. It prepares the reader's mindset and helps them digest what's in the chapter at a later stage. If you are a student writing a thesis then good chapter titles are also important because they demonstrate your writing intentions.

Within chapters you present your sections in a similar way, by carefully breaking down what's in the chapter, by carefully arranging the order, and by carefully providing proper section titles. And so on.

1.3.1 *Minor Document Divisions*

LATEX provides the following sectional units:

part An optional unit that is used for major divisions.
chapter A chapter in a book or report.
sections A section, subsection, or subsubsection.
paragraphs A named paragraph or a named subparagraph.

None of these sectional units are available in the letter class. Using the command starts the sectional unit and defines its the title. The following shows how to define a chapter called *Foundations* and a section called *Notation*. The remaining commands work the same.

```
\chapter{Foundations}                              LATEX Usage
    \section{Notation}
```

When LATEX processes your document it numbers the sectional units. The default mode is to output the numbers before the titles. For example, this section, which has the title *Document Hierarchy*, has the number 1.3. LATEX also supplies *starred* versions of the sectional commands. These commands suppress the numbers of the sectional units. They are called starred versions because their names end in an asterisk (*). The following is an example of the starred versions of the \chapter and \section commands.

```
\chapter*{Main Theorems}                           LATEX Usage
    \section*{A Useful Lemma}
```

An optional argument replaces the sectional unit title in the table of contents. This is useful if the real title is long.

```
\chapter[Wales]%                                   LATEX Usage
        {My Amazingly Amusing Adventures in
         Llanfairpwllgwyngyllgogerychw%
         yrndrobwllllantysiliogogogoch}
```

Figure 1.3
Coarse document divisions

```
\documentclass[12pt,a4paper]{book}
\begin{document}
    \frontmatter
        \maketitle
        \tableofcontents
    \mainmatter
        \chapter{Introduction}
        \chapter{Conclusion}
    \backmatter
        \chapter*{Acknowledgement}
        \addcontentsline{toc}{chapter}{\bibname}
        \bibliography{db}
\end{document}
```

1.3.2 *Major Document Divisions*

Books and theses typically have *front matter, main matter,* and *back matter.* Usually you need the book class to typeset them but institutions such as universities may have their own document class for theses. Some journal or conference article styles also require front, main, and back matter. The following is based on [Lamport 1994, page 80].

front matter Main information about the document: a half and main title page, copyright page, preface or foreword, table of contents, ...
main matter The main body of the document.
back matter Further information about the document and other sources of information: index, afterword, bibliography, acknowledgements, colophon, ...

The commands \frontmatter, \mainmatter, and \backmatter indicate the start of the front, main, and back matter. The artificial example in Figure 1.3 shows how they may be used. Notice that the example does not include any text. In the example, the command \bibliography inserts the bibliography. The command is explained in Section 1.7. The command \addcontentsline inserts an entry for the bibliography in the table of contents.

Notice that the layout of the LaTeX program is such that it gives you a good overview of the structure of the program.

Intermezzo. If you are writing a thesis then you should consider starting by writing down the chapter titles of your thesis first. Your titles should be good and, most importantly, they should be *self-descriptive*: each chapter title should describe what's in its chapter. Make sure you arrange the titles in the proper order. The order of your chapters should maximise the flow of reading. There should be *no forward referencing,* so previous chapters should not rely on definitions of concepts that are defined in subsequent chapters.

```
We prove the following amazing identity.
% A comment.
+---  3 lines: equation () : A = B\,. --------------------
% Another comment.
```

Figure 1.4
Closed fold in folding editor.
The line starting with +--- is
a closed fold. The open fold is
shown in Figure 1.5.

```
We prove the following amazing identity.
% A comment.
\begin{equation}
    A = B\,.
\end{equation}
% Another comment.
```

Figure 1.5
Open fold in folding editor.
This figure shows the result of
opening the fold in Figure 1.4.

A useful tool in this process is the table of contents. The following is how you use it: (1) open your LATEX source file, (2) add a `\tableofcontents` command at the start of your document body, (3) insert your chapter titles with the `\chapter` command, (4) run `latex` twice (why?), and (5) inspect the table of contents in your document viewer. *Only* when you're happy with your titles and their order should you start writing what is in the chapters. Remember that one of the first things the members of your thesis committee will do is study your table of contents. Better make sure they like it.

Note that you may design your chapters in a similar way. Here you start by putting your section titles in the right order. Writing a thesis like this is just like writing a large program in a top-down fashion and filling in the blanks using stepwise refinement.

1.3.3 *The Appendix*

Some documents have appendices. To indicate the start of the appendix section in your document, you use the `\appendix` command. After that, you use the default commands for starting a sectional unit.

```
\appendix
\chapter{Proof of Main Theorem}
  \section{A Useful Lemma}
```
LATEX Usage

1.4 Document Management

LATEX input files have a tendency to grow rapidly. If you don't add additional structure then you will lose control over the content even more rapidly. The following three solutions help you stay in control:[3]

IDE Use a dedicated LATEX IDE. A good IDE should let you edit an entire sectional unit as a whole, move it around, and so on. A good IDE also should provide a high-level view of the document.

folding editor These are editors that let you define hierarchical folds. A fold works just like a sheet of paper. By folding the fold you hide some of the

---

[3] If you know other solutions then I'd like to learn from you.

Figure 1.6

Using the \includeonly and \include commands. The argument of \includeonly is a list of two file names. The \include commands only includes these two files. The remaining files are not included, which saves time when latex is run.

```
\includeonly{Abstract.tex,MainResults.tex}
\begin{document}
    \include{Abstract.tex}
    \include{Introduction.tex}
    \include{Notation.tex}
    \include{MainResults.tex}
    \include{Conclusion.tex}
\end{document}
```

text. By unfolding the fold or by "entering" a fold you can work on the text that's in the fold. Figure 1.4 shows a closed fold of an equation in the vim editor. The fold contains three hidden lines of text. Figure 1.5 shows the open. Folds may also be created for chapters, sections, and environments but this depends on your editor/IDE.

Folds may be used as follows. At the top level of your LaTeX document you define folds for the top-level sectional units of your document. Within these folds, you define folds for the sub-level sectional units, and so on. By creating folds like this you make the structure of your LaTeX document more explicit, thereby making it easier to maintain your document. For example, re-ordering sectional units is now easy.

files LaTeX has commands for including input from other files. By putting the contents of the top-level sectional units in a separate file, you also make the structure in your document more explicit and makes it much easier to see the structure.

LaTeX provides three commands that are related to file inclusion. The first command is \input. This command does not provide much flexibility and it is used on its own. Basically, \input{⟨file⟩} inserts what's in the file ⟨file⟩ at the "current" position. The two remaining commands are \includeonly and \include. These commands provide more flexibility but they are used in tandem. The command \includeonly{⟨file list⟩} is used in the (document) preamble. It takes one argument, which is a list of the files that may be included further on in the document. To include a file, ⟨file⟩, at a certain position you use the command \include{⟨file⟩} at that position. If ⟨file⟩ is in ⟨file list⟩ then it will be included. Otherwise the file will not be included. You can use the command \include several times and for several files. The advantage of this conditional file inclusion mechanism is that it saves precious time when working on large documents because non-included files require no latex processing.

Figure 1.6 provides an example with several \include commands. The command \includeonly at the top of the example tells LaTeX that only the \include statements for the files Abstract.tex and MainResults.tex should be processed.

1.5 Labels and Cross-references

An important aspect of writing a document is *cross-referencing*, i. e., providing references to sectional units, references to tables and fig-

| ```
\chapter{Introduction}
  A short conclusion is
    presented in
    Chapter~\ref{TheEnd}.
\chapter{Conclusion}
\label{TheEnd}
``` | **1 Introduction**<br><br>A short conclusion is presented in Chapter 2.<br><br>**2 Conclusion** | **Figure 1.7**<br>Using \label and \ref |

| ```
\chapter{Introduction}
  A short conclusion is
    presented in
    Chapter~\ref{TheEnd}.
  The conclusion starts on
    Page~\pageref{TheEnd}.
\chapter{Conclusion}
\label{TheEnd}
``` | **1 Introduction**<br><br>A short conclusion is presented in Chapter 2. The conclusion starts on Page 1.<br><br>**2 Conclusion** | **Figure 1.8**<br>Using \pageref |

ures, and so on. Needless to say, LATEX provides support for effective cross-referencing with ease. This section explains the basics for cross-referencing sectional units. The mechanism is the same for cross-referencing figures, tables, theorems, and other notions. Note that this section does not study citations. Citations are studied in Section 1.7.

The basic commands for cross-referencing are the commands \label and \ref.

\label{⟨label⟩}

This defines a logical label, ⟨label⟩, and associates the label with the current environment, i. e., the environment that the \label command is in. At the top level, the environment is the current sectional unit. Inside a given theorem environment it is that theorem environment, inside a given figure environment it is that figure environment, and so on. Once defined, the logical label becomes a handle, which you may use to reference "its" environment. The argument of the \label command may be any sequence of "normal" characters. The only restriction is that the sequence be unique. ☑

\ref{⟨label⟩}

This command substitutes the number of the environment of the label ⟨label⟩. For example, if ⟨label⟩ is the label of the second chapter then \ref{⟨label⟩} results in 2, if ⟨label⟩ is the label of the third section in Chapter 1 then \ref{⟨label⟩} results in 1.3, and so on. ☑

Figure 1.7 shows how to use the \label and the \ref commands. In this example, the tilde symbol (~) prevents LATEX from putting a line-break after the word 'Section.' In effect it *ties* the word 'Section' and the number that is generated by the \ref command. Tieing text is studied in more detail in Section 2.1.1.

The command \pageref{⟨label⟩} substitutes the page number "of" the environment of ⟨label⟩. Figure 1.8 demonstrates how to use the \pageref command.

When you compile a document that references an undefined label, latex will notice this error, complain about it in the form of a warning message, but tacitly ignore the error. Furthermore, it will put two question marks in the output document. The position of the question marks corresponds to the position in the input that references the label. The question marks are typeset in a bold face font: **??**. Even if you fail to notice the warning message this still makes it possible to detect the error.

It should be clear that properly dealing with newly defined or deleted labels requires some form of two-pass system. The first pass detects the label definitions and the second pass inserts the numbers of the labels. When Lamport designed LaTeX he decided that a two-pass system was too time consuming. Instead he decided to compromise:

○ Label definitions are processed by writing them to the auxiliary file for the *next* session.
○ Label references are only considered valid when the labels are defined in the auxiliary file of the *current* session.
○ If an error occurs, information about labels may not be written to the auxiliary file.

Note that with this mechanism latex cannot know about newly defined labels even if a label is referenced after the definition of the label. This is a common cause of confusion. For example, when latex processes a reference to a label that is not defined in the current auxiliary file, it always outputs a message warning about new or undefined labels. The warning is output even if the current input file defines the label. In addition latex will put two question marks where the label is referenced in the text. To the novice user it may seem that they or latex have made an error. Running latex once more usually gets rid of the warnings and question marks.

1.6 Controlling the Style of References

The labelling mechanism is elegant and easy to use but you may still run into problems from a document management perspective. For example, in our previous example, we wrote Chapter~\ref{TheEnd}, thereby hard-coding the word Chapter. If for some reason we decided to change Chapter to Chap for chapter references then we would have to make a change for each command that references a chapter in our source document.

To overcome these problems, and for consistent referencing, it is better to use the prettyref package. Using this package adds a bit of intelligence to the cross-referencing mechanism. There are four ingredients to the new cross-referencing mechanism.

1. You introduce classes of elements. Within each class the elements should be referenced in the same way. For example, the class of equations, the class of figures, the class of chapters, the class consisting of sections, subsections, and subsubsections, and so on.

```
\usepackage{prettyref}
\newrefformat{ch}{Chapter~\ref{#1}}
\newrefformat{sec}{Section~\ref{#1}}
\newrefformat{fig}{Figure~\ref{#1}}
\begin{document}
    \chapter{Introduction}
        In \prettyref{ch:Main@Results}
         we present the main results.
    \chapter{Main Results}
    \label{ch:Main@Results}
    ...
\end{document}
```

Figure 1.9

Using the prettyref package. The \newrefformat commands define three classes of labels: ch, sec, and fig. The command \newrefformat has two arguments. The first argument determines the class and the second determines how the command typesets labels from that class. For example, labels starting with ch: are typeset as Chapter followed by the number of the label.

2. You choose a unique prefix for the labels of the classes. For example, eq for equations, fig for figures, ch for chapters, sec for sections, subsections, and subsubsections, and so on. Here the prefixes are the first few letters of the class names but this is not required.

3. You use the \newrefformat command to specify how each class should be referenced. You do this by telling the command about the unique prefix of the class and the text that should be used for the reference. For example, \newrefformat{ch}{Chapter~\ref{#1}} states that the unique label prefix ch is for a class of elements that have references of the form Chapter~\ref{#1}, where #1 is the logical label of the element (including the prefix). As another example, \newrefformat{id}{\ref{#1}} gives you the same reference you get it you apply \ref to the label.

4. You use \prettyref instead of \ref. This time the labels are of the form ⟨prefix⟩:⟨label⟩. For example, \prettyref{fig:fractal} or \prettyref{ch:Introduction}.

Changing the style of the cross-references of a class now only requires changing a single call to \newrefformat. Clearly, this is much better than having to search for and replace all command instances that label actual chapters. Since prettyref is a package, it should be included in the preamble of your LATEX document. Figure 1.9 provides a complete example of the prettyref mechanism.

1.7 The Bibliography

Most scholarly works have citations and a bibliography or reference section. Throughout this book we shall use the word bibliography for bibliography as well as for reference section. The purpose of the bibliography is to provide details of the works that are cited in the text. We shall refer to cited works as *references*. The purpose of the citation is to acknowledge the reference and to inform your readers how to find the work. In computer science the bibliography is usually at the end of the work. However, in some scientific communities it is common practice to have a bibliography at the end of each chapter in a book.

Figure 1.10
A minimal bibliography

[**Lamport, 1994**] L. Lamport. LaTeX: A Document Preparation System. Addison–Wesley, 1994.

[**Knuth, 1990**] D. E. Knuth. The TeXbook. Addison–Wesley, 1990. The source of this book is freely available from `http://www.ctan.org/tex-archive/systems/knuth/tex/`.

Other communities (e.g., history) use *note systems*. These systems use numbers in the text that refer to footnotes or to notes at the end of the chapter, paper, or book.

The bibliography entries are listed as ⟨citation label⟩ ⟨bibliography content⟩. The ⟨citation label⟩ of a reference is also used when the work is cited in the text. The ⟨bibliography content⟩ lists the relevant information about the work. Figure 1.10 presents an example of two entries in a bibliography. For this example, the citation labels are typeset in boldface font inside square brackets. The `biblatex` package, which is discussed further on, doesn't like the `\,` command in the example. However, replacing this command with `\thinspace` should work.

Even within a single work there may be different styles of citations. *Parenthetical citations* are usually formed by putting one or several citation labels inside square brackets or parentheses. However, there are also other forms of citations that are derived from the information in the citation label.

Within one single bibliography the bibliography content of different kinds of works may differ. For example, entries of journal articles have page numbers but book entries do not.

Bibliographies in different works may also differ. They may have different kinds of (citation) labels and different information in the bibliography content. The order of presentation of the entries in the bibliographies may also be different. For example, entries may be listed alphabetically, in the order of first citation in the text, ...

In LaTeX the style of the bibliography and labels is configurable. Labels may appear as:

numbers This style results in citations that appear as '[⟨number⟩]' in the text.

names and years This style results in citations that appear as '[⟨name⟩, ⟨year⟩]' in the text.

...

Before continuing let's briefly compare labels as numbers and labels as names and years. We shall start with labels as numbers.

Labels as numbers are very compact. They don't disrupt the "flow of reading:" they're easy to skip. Unfortunately, labels as numbers are not very informative. You have to look up which work corresponds to the number in the bibliography. This is annoying because it hinders the reading process. What is worse, labels as numbers lack so much information content that you may have to look up the same number

```
The \LaTeX{} package was
 created by Leslie Lamport%
 ~\cite{Lamport:94}
 on top of Donald Knuth's
 \TeX{} program%
 ~\cite{Knuth:1990}.
```

The LATEX package was created by Leslie Lamport [Lamport 1994] on top of Donald Knuth's TEX program [Knuth 1990].

Figure 1.11
The \cite command

several times. Hyperlinks in electronic documents somewhat reduce this problem.

Labels as names and year are longer than labels as numbers. They are more disruptive to the reading process: they are more difficult to "skip." However, labels as names and years are more informative. If you're familiar with the literature then usually there's no need to go to the bibliography to look up the label. Even if you have to look up which work corresponds to a label you will probably remember it the next time you see the label. Compared to labels as numbers this is a great advantage.

Traditionally, labels for citations appeared as numbers in the text. The main reason for doing this was probably to keep the printing costs low. Nowadays, printing costs are not always relevant.[4] For example, paper is not as expensive as before. Also documents and even journals are distributed electronically. Some journals and universities require specific bibliography styles/formats.

The \bibliographystyle command tells LATEX which style to use for the bibliography. The bibliography style called ⟨style⟩, is defined in the file ⟨style⟩.bst. The following demonstrates how you use the \bibliographystyle command to select a bibliography style called named, which is similar to the style in this book. Though this is not required, it is arguably a good idea to put the \bibliographystyle command in the preamble of your document. The bibliography style named requires the additional package named, which explains why the additional command \usepackage{named} is used in the example.

```
                                                    LATEX Usage
\bibliographystyle{named}
\usepackage{named}
```

The following are a few commonly used bibliography styles. This list is based on [Lamport 1994, pages 70–71].

plain Entries are sorted alphabetically. Labels appear as numbers in the text.
alpha Entries are sorted alphabetically. Labels are formed from surnames and year of publication (e.g., Knut66).
abbrv Entries are very compact and sorted alphabetically. Labels appear as numbers in the text.

Citing a work in LATEX is similar to referencing a section. Both mechanisms use logical labels. For referencing you use the \ref com-

---

[4]But we should think about the environmental effects of using more paper than necessary.

Figure 1.12

Using \cite with an optional argument

```
More information about the
bibliography database
may be found in%
~\cite[Appendix~B]
     {Lamport:94}.
```

More information about the bibliography database may be found in [Lamport 1994, Appendix B].

mand but for citations you use the \cite command. The argument of the \cite command is the logical label of the work you cite.

Figure 1.11 provides an example. The example involves two logical labels: Lamport:94 and Knuth:1990. Each of them is associated with a work in the bibliography. The first label is the logical label of a book by Lamport; the second that of a book by Knuth. As it happens, the names of the labels are similar to the resulting citation labels but this is not required. The command \cite{Lamport:94} results in the citation [Lamport 1994]. Here 'Lamport, 1994' is the citation label of Lamport's book in the bibliography. This label is automatically generated by the BibTeX program. This is explained further on.

The reason for putting an empty group (the two braces) after the \LaTeX and \TeX commands is technical. The following explains this in more detail. As we've seen before, words are separated using inter-word spaces. A group is treated as a word. Most commands are not treaded as words. For example, writing \LaTeX package results in 'LaTeXpackage.' Without the empty groups, there would not have been proper inter-word spacing between 'LaTeX' and 'package' and between 'TeX' and 'program' in the final output. The empty groups after the commands provide the proper inter-word spacing.

It may not be immediately obvious, but in the example of Figure 1.11 the text Lamport on Line 2 in the input is still tied to the command \cite{Lamport:94} on the following line. The reason is that the comment following the text Lamport makes LaTeX ignore all input until the next non-space character on the next line.

You can also cite parts of a work, for example a chapter or a figure. The parts are specified by the optional argument of the \cite command. The example in Figure 1.12 demonstrates how you do this.

The following commands are also related to the bibliography.

\refname

This results in the name of the bibliography section. In the article class, the command \refname is initially defined as 'References.' ☑

\renewcommand\refname{⟨other name⟩}

This redefines the command \refname to ⟨other name⟩. The \renewcommand may also be used to redefine other existing commands. It is explained in Chapter 11. ☑

\nocite{⟨list⟩}

This produces no text but writes the entries in the comma-separated list ⟨list⟩ to the bibliography file. If you use this command, then you should consider making it very clear which works in the bibliography are not cited in the text. For example, some readers may be interested in a discussion of these uncited works, why they are relevant, and so on. They may get very frustrated if they can't find references to these works in your text. ☑

1.7.1 *The bibtex Program*

Since bibliographies are important and since it's easy to get them wrong, some of the work related to the creation of the bibliography has been automated. This is done by BibTeX. The BibTeX tool requires an external human-readable bibliography database. The database may be viewed as a collection of records. Each record defines a work that may be listed in the bibliography. The record defines the title of the work, the author(s) of the work, the year of publication, and so on. The record also defines the logical label of the work. This is the label you use when you \cite the work.

The advantage of using BibTeX is that you provide the information of the entries in the bibliography and that BibTeX generates the text for the bibliography. This guarantees consistency and ease of mind. Furthermore, the BibTeX database is reusable and you may find BibTeX descriptions of many scholarly works on the web. A good starting point is http://citeseer.ist.psu.edu/.

Generating the bibliography with BibTeX is a multi-stage process, which requires an external program called bibtex. The bibtex program is to BibTeX what the latex program is to LATEX. The following explains the process.

1. You specify the name of the BibTeX database and the location of the bibliography with the \bibliography command. The argument of the command is the basename of the BibTeX database. So if you use \bibliography{⟨db⟩}, then your database is ⟨db⟩.bib.
2. You \cite works in your LATEX program. Your logical labels should be defined by some BibTeX record.
3. You run latex. This writes the logical labels to an auxiliary file.
4. You run bibtex as follows:

```
$ bibtex ⟨document⟩
```
Unix Usage

Here ⟨document⟩ is the basename of your top-level LATEX document. The bibtex program will pick up the logical labels from the auxiliary file, look up the corresponding records in the BibTeX database, and generate the code for LATEX's bibliography. A common mistake of bibtex users is that they add the tex extension to the basename of the LATEX source file. It is not clear why this is not allowed.
5. You run latex twice (why?) and Bob's your uncle.

It is important to understand that you (may) have to run bibtex when (1) new citation labels are added, when (2) existing citation labels are removed, and when (3) you change the BibTeX records of works in your bibliography. Each time you run bibtex should be followed by two more latex runs.

Figure 1.13

Including a bibliography

```
\documentclass[11pt]{article}
% Use bibliography style named.
% Requires the file named.bst.
\bibliographystyle{named}
% Requires the package named.sty.
\usepackage{named}
\begin{document}
    % Put in a citation.
    This cites~\cite{Knuth:1990}.
    % Put the reference section here.
    % It is in the file db.bib.
    \bibliography{db}
\end{document}
```

Figure 1.14

Some BIBTEX entries

```
@Book{Lamport:94,
   author   = {Lamport, Leslie},
   title    = {\LaTeX: A Document Preparation System},
   year     = {1994},
   isbn     = {0-021-52983-1},
   publisher = {Addison\,\endash\,Wesley},
}

@Book{Strunk:White,
   author   = {Strunk, W. and
                 White, E.\,B.},
   title    = {The Elements of Style},
   publisher = {Macmillan Publishing},
   year     = {1979},
}
```

Figure 1.13 provides an example. The LaTeX source in this example depends on a BibTeX file called db.bib.

Figure 1.14 is an example of two entries in a BibTeX database file. The example associates the logical label Lamport:94 with Lamport's LaTeX book and the logical label Strunk:White with the book about elements of style. The entries also specify the author, title, year of publication, International Standard Book Number (ISBN), and the publisher. Some of this information is redundant: this depends on the style that is used to generate the bibliography. Note that you first provide an author's surname and then the first name(s), separating them with a comma. The second entry in the example shows that multiple authors are separated with the keyword and. The \endash in the first publisher inserts the correct dash symbol.

Now that you know how to use the bibtex program, let's see what you can put in the BibTeX database. The following list is not exhaustive. For a more accurate list you may wish to read [Fenn 2006; Lamport 1994]. The following is based on [Lamport 1994, Appendix B].

```
\usepackage[style=authoryear,
            block=space,
            language=british]{biblatex}
\renewcommand*\bibopenparen{[}
\renewcommand*\bibcloseparen{]}
\renewcommand*\bibnamedash
            {\rule[0.48ex]{3em}{0.14ex}\space}
\addbibresource{LAF}
```

Figure 1.15

Using biblatex

@Article An article from a journal or magazine.
> **required entries** author, title, journal, and year.
> **optional entries** volume, number, pages, month, and note.

@Book A book with an explicit publisher.
> **required entries** author or editor, title, publisher, and year.
> **optional entries** volume, number, series, ...

@InProceedings A paper in a conference proceedings.
> **required entries** author, title, booktitle, publisher, and year.
> **optional entries** pages, editor, volume, number, series, ...

@Proceedings The proceedings of a conference.
> **required entries** title and year.
> **optional entries** editor, volume, number, series, organisation,
> ....

@MastersThesis A Master's thesis.
> **required entries** author, title, school, and year.
> **optional entries** type, address, month, and note.

@PhDThesis A Ph.D. thesis.
> **required entries** author, title, school, and year.
> **optional entries** type, address, month, and note.

....

Turner [2010] presents an impressive list of BibTeX style examples.

1.7.2 *The biblatex Package*

There are several problems with the basic LATEX citation mechanism. The biblatex package overcomes many of these. It provides a more flexible citation mechanism and lets you configure your own citation style. The biblatex package is very comprehensive. Is it beyond the scope of this book to discuss it in detail. In the remainder of this section we shall explore some of its capabilities.

The bibliography management for this book was created with biblatex. Figure 1.15 shows how this was done. First we load the biblatex package. Using the options of the \usepackage we set the citation style to author year. The other options are less important. The first two \renewcommand* commands redefine the symbols that are used for citations inside parentheses/brackets. The \renewcommand* command is explained in Chapter 11. For this book, the opening bracket is a square bracket and the closing bracket a closing square

Figure 1.16

Textual and parenthetical citations

```
\textcite{Knuth:1990}
  describes \TeX.
The ultimate guide to \TeX{}
  is~\parencite{Knuth:1990}.
```

Knuth [1990] describes TeX. The ultimate guide to TeX is [Knuth, 1990].

Figure 1.17

Getting the author and year of a citation

```
\citeauthor{Knuth:1990}
wrote~\parencite{Knuth:1990}
in~\citeyear{Knuth:1990}.
```

Knuth wrote [Knuth, 1990] in 1990.

bracket. The third `\renewcommand*` redefines the command that produces the symbol that is used for repeated author names in entries in the bibliography. In this book the symbol is a horizontal rule to the length of 3 em, which is followed by a single space. The 3 em long rule is recommended by Bringhurst [2008, page 80]. The last command specifies the name of the bibliography database file.

To print the bibliography, you use the `\printbibliography` command. The command takes an optional argument that lets you specify the title of the bibliography. The optional argument may also be used as a filter for what entries should be admitted in the bibliography. Some of this is explained in Sections 1.7.3 and 1.7.4.

The following shows how you print the bibliography of your work. The title of the bibliography section is set to 'References.'

```
\printbibliography[title=References]
```
LaTeX Usage

○ The `biblatex` package distinguishes between *parenthetical* and *textual* citations. A *parenthetical* citation is like LaTeX's default citation. Such citations are not supposed to play any rôle at the sentence level. You get parenthetical citations with the `\parencite` command: [Knuth 1990]. A *textual* citation acts as a subject or an object in a sentence: Knuth [1990]. Leaving out such citations should result in grammatical errors. You get textual citations with the `\textcite` command. Figure 1.16 demonstrates how these commands work.

○ The `biblatex` package also provides two commands called `\citeauthor` and `\citeyear`. The first command gives you the author(s) and the second the year of a citation. Figure 1.17 shows how to use these commands. There is also a `\citetitle` command, which works as expected.

○ An important improvement is that `biblatex` lets you capitalise "von" parts in surnames. To achieve this you use similar commands as before. However, this time the relevant commands start with an uppercase letter. The example in Figure 1.18 demonstrates how this works.

The `biblatex` also supports footnotes and endnotes but that is not explained in this book. The interested reader may find more information about these features in the package documentation [*The biblatex Package*]. In linux you may get package information about the `biblatex` package by executing the following command.

```
\Citeauthor{Beethoven:ninth}
 is most famous for his Ninth Symphony%
 ~\Parencite{Beethoven:ninth}.
Personally, I prefer his Sixth Symphony%
 ~\Parencite{Beethoven:sixth}.
```

Figure 1.18

Using biblatex's citation commands

```
$ texdoc biblatex
```
Unix Usage

Getting help for other packages and classes works similarly. If your LATEX distribution doesn't come with biblatex then you may download it from the Comprehensive TEX Archive Network (CTAN), which may be found at http://www.ctan.org. If you are looking for other classes and packages then CTAN is also the place to be.

1.7.3 *End-of-Chapter Bibliographies*

Some documents require a bibliography at the end of each section or chapter. For example, in a conference proceedings each article has its own bibliography. Some theses have a separate bibliography for each chapter or part. The remainder of this section shows how to get a bibliography at the end of a chapter with the aide of biblatex.

1. You import biblatex with your favourite options.

```
\usepackage[⟨options⟩]{biblatex}
```
LATEX Usage

2. You specify the names of your bibliography database(s).

```
\addbibresource{⟨your .bib file names⟩}
```
LATEX Usage

3. You add a refsection environment for each chapter and print the bibliography at the end of the chapter. The following shows how this is done for one chapter—of course you can repeat this for other chapters. The option heading=subbibliography prints the default name of the bibliography in the same style as that of a section.

```
\chapter{From K\"onigsberg to G\"ottingen}
\begin{refsection}
   ... % Lots of text and citations omitted.
\printbibliography[heading=subbibliography]
\end{refsection}
```
LATEX Usage

4. You run latex on your LATEX source file. This will create an auxiliary file for each refsection with a \printbibligraphy command in it. The names of theses auxiliary files are of the form ⟨base name⟩⟨number⟩-blx.aux, where ⟨base name⟩ is the base name of your main document.

5. You run bibtex on each auxiliary file. In unix this may be done as follows.

```
$ for f in *[0-9]-blx.aux; do biblatex $f; done
```
Unix Usage

6. You run LATEX twice.

7. You sit down, relax, and admire your end-of-chapter bibliographies.

1.7.4 *Classified Bibliographies*

This section explains how you create documents with *classified bibliographies*—Lehman [*The biblatex Package*] calls them *subdivided bibliographies*. The technique in this section can also be used to create end-of-chapter bibliographies but you would be silly if you did because it would take more time. Classified bibliographies are quite common. For example, some works have bibliographies for books, for journal articles, for conference papers, and so on.

As in the previous two sections biblatex comes to our rescue. This time, we have to work a bit harder because biblatex cannot automatically recognise the categories. We shall start with a slight variation on the end-of-chapter bibliographies. For simplicity the details about loading and configuring biblatex are omitted.

Let's assume you want one bibliography section in your main document but you want separate subbiliographies for chapters.

1. You add refsection environments for your chapters.

```
                                                         LaTeX Usage
\chapter{Philip Glass}
\begin{refsection}
  ... % lots of text and citations omitted.
\end{refsection}
% Steve Reich, John Adams and Arvo Pärt chapters omitted.
```

2. The following prints a title for the collected subbibliographies.

```
                                                         LaTeX Usage
\printbibheading
```

By default the previous command results in the text 'References' but you can configure it.

```
                                                         LaTeX Usage
\defbibheading[heading=bibliography,
                title=Classified Discographies]
```

3. You print the subbibliographies.

```
...                                                      LaTeX Usage
\printbibliography[section=1,title=Glass Discography]
\printbibliography[section=2,title=Reich Discography]
...
```

The section numbers 1 and 2 refer to the order of appearance of the refsection environments. The 1 corresponds to the first environment/chapter and the 2 to the second environment/chapter. This is a bit unfortunate, because it means that if you rearrange the two environments then you have to change the numbers in the \printbibliography commands.

4. You complete the process by running LaTeX, running BibTeX on the auxiliary files and running LaTeX two more times.

Needless to say, you're quite lucky if your subbibliographies correspond to your chapters. In general this is not possible. For example you may want a bibliography for books and one for articles. Again `biblatex` comes to the rescue because it can separate bibliographies by the type of BIBTEX entry: `@Book`, `@Article`, ... The following shows how to create your subbibliographies.

```
\printbibliography[type=book,title=Books]                    LATEX Usage
\printbibliography[type=article,title=Journal Articles]
```

In general the presented techniques are not always sufficient, which is why `biblatex` allows *keywords* in your BIBTEX entries. As a matter of fact, the package also introduces a whole range of non-standard BIBTEX entries such as `@Online`, `@Patent`, `@Reference`, and so on. The following shows how you associate three keywords called `glass`, `opera`, and `minimal` with a `@Misc` entry of an opera by Philip Glass.

```
@Misc{Akhnaten,                                              BIBTEX
   title    = {Akhnaten},
   author   = {Glass, Philip},
   keywords = {glass,opera,minimal},
   year     = {1983},
}
```

Having defined the keys, you can use them as criteria for your classified bibliographies.

```
\printbibliography[heading=subbibliography,                  LATEX Usage
                   title=Opera References,keyword=opera]
```

A more general concept is that of a named *category*, which you can add bibliography labels to:

```
\DeclareBibliographyCategory{trilogy}                        LATEX Usage
\addtocategory{trilogy}{Akhnaten,Einstein,Satyagraha}
```

Having defined the category, you can use it to generate a subbibliography.

```
\printbibliography[heading=subbibliography,                  LATEX Usage
                   title=Trilogy References,
                   category=trilogy]
```

1.8 Table of Contents and Lists of Things

This section explains how to include a table of contents and *reference lists* in your document. Here a reference list is a list that tells you where (in the document) you may find certain things. Common examples are a list of figures, a list of tables, and so on. LATEX also lets you define other reference lists. The example in Figure 1.19 shows how you include a table of contents, and lists of figures and tables.

In the example, the `\clearpage` command inserts a pagebreak after the first `\include` command. As a side-effect it also forces any figures

Figure 1.19
Including reference lists

```
\begin{document}
    \maketitle
    \include{Abstract.tex}
    \clearpage
    \tableofcontents
    \listoffigures
    \listoftables
    ⋮
\end{document}
```

and tables that have so far appeared in the input to be printed. There is also a command called \cleardoublepage, which works similarly. However, in a two-sided printing style, it also makes the next page a right-hand side (odd) page, producing a blank page if necessary.

1.8.1 *Controlling the Table of Contents*

The *counter* \tocdepth (counters are discussed in Chapter 12) gives some control over what is listed in the table of contents. The value of the counter controls the depth of last sectional level that is listed in the table of contents. The value 0 represents the highest sectional unit, 1 the next sectional unit, and so on.

By setting the value of \tocdepth to ⟨depth⟩ you limit the depths of the sectional units that are listed in the table of contents from 0 to ⟨depth⟩. For example, if you're using the book class, then using 0 for ⟨depth⟩ will allow parts and chapters in the table of contents, but not sections. As another example, if you're using the article class, then using 2 for ⟨depth⟩ will only list sections, subsections, and subsubsections in the table of contents. You set the counter \tocdepth to ⟨depth⟩ with the command \setcounter{\tocdepth}{⟨depth⟩}.

1.8.2 *Controlling the Sectional Unit Numbering*

The counter \secnumdepth is related to the counter \tocdepth. Its value determines the depth of the the sectional units that are numbered. So by setting the counter \secnumdepth to 3, you tell LATEX to number parts, chapter, and sections, and tell it to stop numbering subsections and less significant sectional units.

Table 1.1 lists the sectional unit commands and the corresponding numbers for the counters \tocdepth and \secnumdepth.

1.8.3 *Indexes and Glossaries*

If you are writing a book or a thesis, you probably want to include an index or glossary of some kind. Getting it to work may take a while. The remainder of this section explains how to create an index. The mechanism for glossaries is similar.

Unfortunately, LATEX's default index mechanism only allows you

Sectional Unit Command	\tocdepth	\secnumdepth
\part	-1	1
\chapter	0	2
\section	1	3
\subsection	2	4
\subsubsection	3	5
\paragraph	4	6
\subparagraph	5	7

Table 1.1
Depth values of sectional unit commands. The first column in the table lists the sectional unit commands. For each command, the corresponding value of the \tocdepth counter is listed in the second column. That of the \secnumdepth counter value is listed in the last column.

to have one single index. The multind package lets you create several index lists. The package works as follows.

1. You associate each index with a file name. You do this by passing the basename of the file to the command \makeindex.

```
\makeindex{programs}                                    LATEX Usage
\makeindex{authors}
```

2. You insert the indexes with the \printindex command.

```
\printindex{programs}{Index of Programs}                LATEX Usage
\printindex{authors}{Index of Authors}
```

The first argument of \printindex is the name of the corresponding index. The second name is the title of the index. The title also appears in the table of contents.

3. You define the index entries. You use the \index command to define what is in the indexes. The following is a simple example that creates an entry 'TeX' in the index for the programs.

```
Knuth\index{authors}{Knuth}                             LATEX Usage
   is the author of \TeX\index{programs}{TeX}.
```

Behind the scenes the \index command writes information to the auxiliary files authors.idx and programs.idx. In the following step we shall use the makeindex program to turn it into files that can be included in our final document.

4. You process the idx files with the program makeindex. This is similar to using bibtex for generating the bibliography. The following demonstrates how to use the program.

```
$ makeindex authors                                     Unix Session
$ makeindex programs
```

The remainder of this section explains how you create more complex indexes with the multind package. The multind package redefines the \index command. The redefined command takes one more argument. The first argument of the redefined command determines the name of an auxiliary file that is used to construct the index. The last argument of the redefined command describes the index entry. The following is based on [Lamport 1994, Appendix A.2].

`\index{⟨name⟩}{⟨entry⟩}`

> This creates an index entry for ⟨entry⟩. The entry also lists the page number. ☑

`\index{⟨name⟩}{⟨entry⟩!⟨subentry⟩}`

> This creates a subentry ⟨subentry⟩ for the index entry ⟨entry⟩. It also lists the page number. ☑

`\index{⟨name⟩}{⟨entry⟩!⟨subentry⟩!⟨subsubentry⟩}`

> This creates a sub-subentry ⟨subsubentry⟩ for subentry ⟨subentry⟩ for the index entry ⟨entry⟩. It also lists the page number. ☑

`\index{⟨name⟩}{⟨entry⟩|see{⟨other entry⟩}}`

> Creates a cross-reference. in the entry for ⟨entry⟩. This does not result in a page number for ⟨entry⟩. ☑

`\index{⟨name⟩}{⟨entry for sorting⟩@{⟨entry for printing⟩}}`

> This results in an entry for ⟨entry for printing⟩ in the index list. The position in the index is determined by ⟨entry for sorting⟩. This is useful, for example, if ⟨entry for printing⟩ contains mixed upper- and lowercase letters or if it contains other characters. ☑

> The following are some examples of the last construct.

- ○ `\index{⟨name⟩}{twenty@20};`
- ○ `\index{⟨name⟩}{twenty@xx};`
- ○ `\index{⟨name⟩}{beta@$\beta$};` or
- ○ `\index{⟨name⟩}{command@\texttt{{\textbackslash}command}}.`

There is one more construct, which is useful for topics that cover a page range. To create an entry for topic ⟨topic⟩ for a page range, you start the range with the command `\index{⟨name⟩}{⟨topic⟩|(}` and you close the range with the command `\index{⟨name⟩}{⟨topic⟩|)}`.

Table 1.2 presents an example of the `\index` command. The left of the figure depicts the last argument of the `\index` commands and the current page of the LaTeX output. It is assumed that the first argument of the `\index` command is the same. The right of the figure depicts the resulting index. Notice that the entries for programs and sausages both have subentries. However, the entry for programs has a page number whereas the entry for sausages does not. To understand this difference you need to know that the page number for top-level entries is generated by commands of the form `\index{⟨name⟩}{⟨entry⟩}`. Since there is such a command for sausages but not for programs this explains the difference. More information about the `\index` command may be found in [Lamport 1994, Appendix A].

1.9 Class Files

As explained before, each top-level LaTeX document corresponds to a document class. The document class is determined by the required argument of `\documentclass` command in your LaTeX document.

Page	Last argument of the \index command
1	lecture notes
2	programs
4	lard
2	latex@\LaTeX
3	lambda@λ
5	sausages!boerewors
6	sausages!salami
2	programs!latex
6	programs!bibtex
2	index|(
6	index|)
8	salami|see{sausages}
8	boerewors|see{sausages}
8	boereworst (Dutch)|see{boerewors}

Index

boerewors, *see* sausages
boereworst (Dutch), *see* boerewors

index, 2–6

λ, 3
lard, 4
LATEX, 2
lecture notes, 1

programs, 2
    bibtex, 6
    latex, 2

salami, *see* sausages
sausages
    boerewors, 5
    salami, 6

Table 1.2
Using the \index command. The table on the left lists the last argument of the \index command and the corresponding page in the output document. So if you're using the unmodified \index command then the second column in the table corresponds to the first argument of \index. However, if you're using the multind package then the column corresponds to the command's second argument. The output on the right is the resulting index. The font in the output may seem tiny, but it is representative for a typical index.

```
\documentclass{⟨document class name⟩}                    LATEX Usage
```

Each document class is defined in a class file. Class files define the general rules for typesetting the document. Recall that you can pass options to classes. This is done by putting the options inside the square brackets following the command \documentclass. If you have multiple options then you separate them with commas.

The extension of class files is cls. The following are some standard classes.

article The basic article class. The top-level sectional unit of this class is the section.

book The basic book class. The top-level sectional unit of this class is the chapter. The book class also provides the commands for indicating the start of the front, main, and back matter.

report The basic report class. The top-level sectional unit of this class is the chapter.

letter The basic class for letters. This class has no sectional units. The letter is written inside a letter environment. The letter environment has one required argument, which should be the address of the addressee. In addition there are commands for specifying the address of the writer, the signature, the opening and closing lines, the "carbon copy" list, and enclosures and postscriptum. More detailed information about the letter class may be found in [Lamport 1994, page 84–86] and on http://en.wikibooks.org/wiki/LaTeX/Letters. Figure 1.20 presents a minimal example of a letter.

The following options are typically available for the previously mentioned classes.

Figure 1.20
Minimal letter

```
\documentclass{letter}
% Sender details.
\signature{Donal}
\address{Collect Cash\\Dublin}

\begin{document}
  % Addressee. A double backslash generates a newline.
  \begin{letter}{Donate Cash\\Cork}
    \opening{Dear Sir/Madam:}

    Please make a cash donation to our party.

    We look forward to the money.

    \closing{Yours Faithfully,}
    \ps{P.S. Send it now.}
    \encl{Empty brown envelope.}
    \cc{Paddy.}
  \end{letter}
\end{document}
```

11pt Use an 11 point font size instead of the 10 point size, which is the default.

12pt Use a 12 point font size.

twoside Output a document that is printed on both sides of the paper.

twocolumn Output a document that has two columns.

draft Used for draft versions. This option makes LaTeX indicate hyphenation and justification problems by putting a little square in the margin of the problem line.

final Used for the final version. This is the opposite of draft.

1.10 Packages

Document classes are fairly minimal. Usually, you need some additional commands for doing your day-to-day document preparation. This is where *packages* (originally called *style files*) come into play. Packages have the following purpose.

provide commands Provide new useful commands. Usually, this adds some extra functionality.

change commands Tweak some existing commands. This may change the default document settings. Usually, this affects the layout.

The extension of packages is sty. You include the package that is defined in the file ⟨style⟩.sty as follows.

```
\usepackage{⟨style⟩}
```
LaTeX Usage

Some packages accept options. You pass them to the package using the same mechanism as with classes. Multiple options are separated using commas. The following shows how to pass the options `draft` and `colorlinks` to the package `hyperref`.

```
\usepackage[draft,colorlinks]{hyperref}
```
LATEX Usage

To find out about the options you have to read the documentation. Remember that `texdoc` (see page 27) helps you locate and read the documentation.

1.11 Useful Classes and Packages

There are hundreds, if not thousands, of existing classes and packages. The following are some useful classes and packages that are not discussed in the remainder of this book.

url Typesets URLs [Arseneau 2010] with automatic line breaking.

fourier Sets the text font to *Utopia Regular* and the math font to *Fourier* [Bovani 2005].

coverpage Facilitates user-defined coverpages [Mühlich 2006].

fancyhdr Facilitates user-defined headers and footers [van Oostrum 2004].

lastpage Defines a command for getting the last page number [Goldberg, and Münch 2011]. This is especially useful for M/N page numbers in combination with the `fancyhdr` package [van Oostrum 2004].

mathdesign This package replaces all math symbols with a complete math font. It can be configured with expert and poor man's versions of fonts such as *Adobe Utopia*, *ITC Garamond*, and *Bitstream Charter*.

memoir This class provides support for writing books. The class comes with lots and lots of options for finetuning the typesetting [Wilson, and Madsen 2011].

todonotes This package supports todo notes in the margin and a list of todo notes with hyperlinks. [Midtiby 2011].

classicthesis Nice package for theses [Miede 2010].

arsclassica Another nice package for theses [Patieri 2010]. It is based on `classicthesis`.

mathtools Provides better typesetting of mathematical content [Høgholm et al. 2011].

1.12 Errors and Troubleshooting

One of LATEX's greatest weaknesses are its error messages, which are not always so easy to understand.

When an error occurs LATEX usually outputs the line number where the error was detected. This should give some idea of where to find the source of the error. If you run LATEX in interactive mode then LATEX will output an error message and waits for you to tell it what to do with the error. This is done by typing in an option. The following are some of the options you can type in. More options may be found in [Lamport 1994, Chapter 8].

x This exits the compilation process.

⟨return⟩ This ignores the error and resumes compilation.

q This also ignores the error and resumes compilation. Future errors will not be reported.

h This asks for more help.

The remainder of this section lists some of the more common errors and, where possible, provides some suggestions on how to fix the errors. More information about errors and troubleshooting may be found in [Lamport 1994] or in the UK TₑX FAQ (`http://www.tex.ac.uk/cgi-bin/texfaq2html`).

One of the more common error messages is the following. It is instructive to study it in more detail.

Undefined control sequence
l.⟨line number⟩ ⟨cmd⟩
This usually means there is a typo in the commmand ⟨cmd⟩. The ⟨line number⟩ is the line number where the error was detected. Most error messages report the line number where the error was detected. In the following the line numbers are omitted.

The following are other common error messages. The error messages are sorted alphabetically and the line number information is omitted.

\begin{⟨name₁⟩} ended by \end{⟨name₂⟩}
This means that there is a \begin{⟨name₁⟩} that isn't properly ended by an \end{⟨name₁⟩}. LᴬTₑX detects this error when it discovers the \end{⟨name₂⟩} in the input.

Command ⟨cmd⟩ already defined
This means that there's an attempt to define the existing command ⟨cmd⟩ with one of LᴬTₑX's commands for defining non-existing commands.

Environment ⟨name⟩ undefined
This means there is a \begin{⟨name⟩} in the input and that the environment ⟨name⟩ isn't defined.

Extra alignment tab has been changed to \cr
This means there is an ampersand (&) too many for a row in a tabular-like environment.

\include cannot be nested
This means that there is an \include command in a file that's included by another \include command, which is not allowed.

LaTeX Error: Option clash for package ⟨name⟩
This is a very nasty error that is thrown by the LᴬTₑX kernel. It means that the package ⟨name⟩ is loaded with a set of options, S_1, and that the package is loaded again with options, S_2, where S_2 contains an option that is not contained in S_1. This commonly happens if you're using the hyperref package.

LaTeX Error: Too many unprocessed floats.

This usually means that LaTeX can't position your floats but it may also mean there are too many \marginpar commands. If the problem is related to floats and if you haven't used the option p in your float placement options then adding it may resolve the problem. Remember that the float placement option defaults to tbph. The \clearpage clears the page and also force all pending floats to be printed but this is may not be a satisfactory solution.

Missing $ inserted

This means that you entered a math command in ordinary text mode.

not in outer par mode

A float ar a \marginpar command is used in math mode or inside a "parbox."

Something's wrong--perhaps a missing \item

This means that you forgot to put an \item command at the start of an itemize, enumerate, or description environment.

There's no line to end here

There's a \\ or a \newline command between paragraphs.

TeX capacity exceeded, sorry

This usually means there's a recursive command that doesn't terminate properly. For example, with the definition \newcommand\nono{\no\no} using \no would lead to this error.

Too many }'s

This means that you entered a closing brace too many.

Too deeply nested!

This means that the nesting level of list environments such as itemize or enumerate is too deep.

Runaway argument?

{⟨actual parameter⟩

This may mean that you forgot to enter a closing brace of a compound parameter. LaTeX outputs the value of the actual parameter, which should give you a clue as to where to search. Also see the following error, which is related.

Runaway argument?

⟨actual parameter⟩

! File ended while scanning use of ⟨cmd⟩

This means that ⟨cmd⟩ is a TeX command with a delimited parameter list. When ⟨cmd⟩ was used with ⟨actual parameter⟩ in the input, TeX couldn't find the the delimiter that should have followed ⟨actual parameter⟩.

Unknown option ⟨option⟩ for ⟨class or package⟩

This means there's an attempt to load ⟨class or package⟩ with the invalid option ⟨option⟩.

PART II

Basic Typesetting

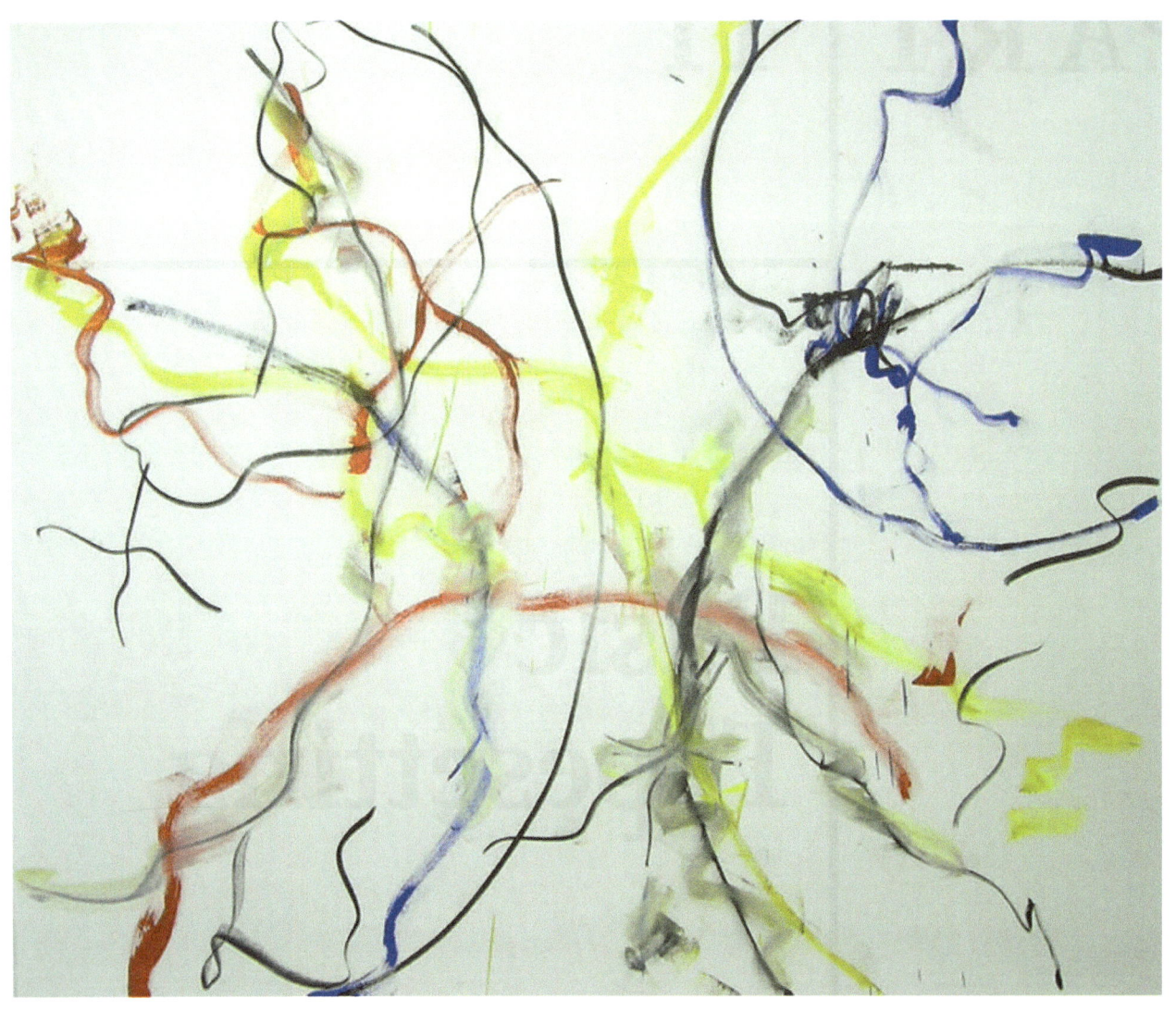

Oil and charcoal on canvas (31/08/05), 183 × 223 cm
Work included courtesy of Billy Foley
© Billy Foley (www.billyfoley.com)

CHAPTER 2
Running Text

THIS CHAPTER EXPLAINS everything you've always wanted to know about writing text, aligning it, and changing text appearance.

Recall from Chapter 1 that LaTeX is implemented on top of TeX, which is a rewriting machine that turns token streams into token streams. Some of the character tokens in the input stream have a special meaning to TeX. This is studied in Section 2.1. The rest of the chapter is about typesetting. We start with some sections about diacritics, ligatures, dashes, emphasis, footnotes and marginal notes, quotes and quotations. If you're not familiar with these notions then don't worry, because they are explained further on. Also you can visit the typography jargon reference on page 267. This chapter ends with sections about changing the size and the type style of the text, the most important text alignment techniques, and language related issues.

2.1 Special Characters

This section studies ten characters that have a special meaning to TeX. When TeX sees these characters as tokens in the input stream, then it usually does not typeset them but, instead, changes state. The remainder of this section briefly explains the purpose of the tokens and how you typeset them as characters in the output.

Table 2.1 depicts the tokens, their meaning, and the command to typeset them. We have already studied the start-of-comment token (%) and the backslash (\), which starts control sequences. Typesetting a backslash is done with the commands `\textbackslash` and `\backslash`. The latter command is only used when specifying mathematical formulae, which is described in Chapter 8. The parameter reference token is described in Chapter 11. The alignment tab (&) is described in Section 2.19.3. This token usually indicates a horizontal alignment position in array-like structures consisting of rows and columns. The math mode switch token ($), the subscript token (_), and the explained token (^) are described in Chapter 8. The three remaining tokens are described in the remainder of this section.

2.1.1 *Tieing Text*

Remember that LaTeX is a large rewriting machine that repeatedly turns token sequences into token sequences. At some stage it turns a

Table 2.1

Token	Purpose	Command
#	parameter reference	\#
$	math mode switch	\$
%	start of comment	\%
&	alignment tab	\&
~	text tie token	\textasciitilde
_	math subscript	_
^	math superscript	\textasciicircum
{	start of group	\{
}	end of group	\}
\	start of command	\textbackslash or \backslash

token sequence into lines. This is where LaTeX (TeX really) determines the line breaks. The tilde token (~) defines an inter-word space that cannot be turned into a line break. As such it may be viewed as an operator that ties words.

The following shows two important applications of the tilde operator: it prevents unpleasant linebreaks in references and citations.

```
...  Figure~\ref{fig:list@format}
  depicts the format of a list.
 It is a reproduction of~\cite[Figure~6.3]{Lamport:94}.
```
LaTeX Usage

It is usually not too difficult to decide where to use the tie operator. The following are some concrete examples, which are taken from [Knuth 1990, Chapter 14].

○ References to named parts of a document:
  ⋆ `Chapter~12,`
  ⋆ `Theorem~1.5,`
  ⋆ ....

Knuth [1990] recommends that you use `Lemmas 5 and~6` because having the 5 at the start of a line is not really a problem.

○ Between a person's forenames and between multiple surnames:
  ⋆ `Donald~E. Knuth,`
  ⋆ `Luis~I. Trabb~Pardo,`
  ⋆ `Bartel~Leendert van~der~Waarden,`
  ⋆ `Charles~XII,`
  ⋆ ....

○ Between math symbols in apposition with nouns:
  ⋆ `dimension~$d$,`
  ⋆ `string~$s$ of length~$1$,`
  ⋆ ....

Here the construct $\langle math \rangle$ is used to typeset $\langle math \rangle$ as an in-line mathematical expression.

○ Between symbols in series:
  ⋆ `1,~2, or~3.`

○ When a symbol is a tightly bound object of a preposition:

```
⋆ from 0 to~1,
⋆ increase $z$ by~1,
⋆ ....
```
○ When mathematical phrases are rendered in words:
```
⋆ equals~$n$,
⋆ less than~$\epsilon$,
⋆ modulo~$2$,
⋆ for large~$n$,
⋆ ....
```
○ When cases are being enumerated within a paragraph:
```
⋆ Show that function $f( x )$ is (1)~continuous; (2)~bounded.
```

2.1.2 *Grouping*

Grouping is a common technique in LaTeX. The opening brace ({) starts a group and closing brace (}) closes it. Grouping has two purposes. The first purpose of grouping is that it turns several things into one compound thing. This may be needed, for example, if you want to pass several words to a command that typesets its argument in bold face text. The following demonstrates the point.

```
A bold \textbf{word} and
a bold \textbf letter.
```
A bold **word** and a bold **l**etter.

The second purpose of grouping is that it lets you change certain settings and keep the changes local to the group. The following demonstrates how this may be used to make a local change to the type style of the text inside the group.

```
Normal text here.
{% Start a group.
 \bfseries
 % Now we have bold text.
 Bold paragraphs in here.
}% Close the group.
Back to normal text again.
```
Normal text here. **Bold paragraphs in here.** Back to normal text again.

Inside the group you may have several paragraphs. The advantage of the declaration \bfseries is that it defines how the text is typeset until the end of the group. The \textbf command just typesets its argument in a bold typeface. The argument may not contain paragraph-breaks.

There is also a low-level TeX mechanism for creating groups. It works just as the braces. A group is started with \begingroup and ended with \endgroup. These tokens may be freely mixed with braces but {/} pairs and \begingroup/\endgroup pairs should be properly matched. So { \begingroup \endgroup } is allowed but { \begingroup } \endgroup is not. A brace pair affects whitespace when you're typesetting mathematics but a \begingroup/\endgroup pair does not.

Table 2.2
Common diacritics

Output	Command	Name
ò	\`{o}	Acute accent
ó	\'{o}	Grave accent
ô	\^{o}	Circumflex (hat)
õ	\~{o}	Tilde (squiggle)
ö	\"{o}	Umlaut or dieresis
ċ	\.{c}	Dot accent
š	\v{s}	Háček (caron or check)
ŏ	\u{o}	Breve accent
ō	\={o}	Macron (bar)
ő	\H{o}	Long Hungarian umlaut
o͡o	\t{oo}	Tie-after accent
ş	\c{s}	Cedilla accent
ọ	\d{o}	Dot-under accent
o̲	\b{o}	Bar-under accent

Table 2.3
Other special characters

Output	Command	Name
å	\aa	Scandinavian a-with-circle
Å	\AA	Scandinavian A-with-circle
ł	\l	Polish suppressed-l
Ł	\L	Polish suppressed-L
ø	\o	Scandinavian o-with-slash
Ø	\O	Scandinavian O-with-slash
¿	?`	Open question mark
¡	!`	Open exclamation mark

2.2 Diacritics

This section studies how to typeset characters with *diacritics*, which are also known as accents. Table 2.2 displays some commonly occurring diacritics and the commands that typeset them. The presentation is based on [Knuth 1990, Chapter 9].

Using \"{i} to typeset ï may not work if you're not using a Type 1 font (T1 font). However, typesetting ï with \"{\i} should always work. Here the command \i is used to typeset a dotless i (ı). There is also a command \j for a dotless j.

Table 2.3 shows some other commonly occurring special characters.

2.3 Ligatures

A ligature combines two or several characters as a special glyph. Examples of English ligatures and their equivalent character combinations are fi (fi), ff (ff), ffi (ffi), fl (fl), and and ffl (ffl). LaTeX recognises English ligatures and substitutes them for the characters representing them.

Table 2.4 displays some foreign ligatures. The symbol ß (eszett) is

Output	Command	Name
œ	\oe	French ligature œ
Œ	\OE	French ligature Œ
æ	\ae	Scandinavian ligature æ
Æ	\AE	Scandinavian ligature Æ
ß	\ss	German 'Eszett' or sharp S

Table 2.4
Foreign ligatures

```
'Convention' dictates that
punctuation go inside
quotes, like ''this,'' but
some think it's better
to do ''this''.
```
'Convention' dictates that punctuation go inside quotes, like "this," but some think it's better to do "this".

Figure 2.1
Quotes

```
''\,'Fi' or 'fum?'\,'' he asked.\\
'''Fi' or 'fum?''' he asked.    \\
''{}'Fi' or 'fum?'{}'' he asked.
```
" 'Fi' or 'fum?' " he asked.
"'Fi' or 'fum?'" he asked.
"'Fi' or 'fum?'" he asked.

Figure 2.2
Nested quotations

a ligature of ſs [Bringhurst 2008] and this is reflected in the LaTeX command that typesets the symbol.

Sometimes it is better to suppress ligatures. The following is an example: the \makebox command prevents LaTeX from turning the fi in selfish into a ligature, which makes the result much easier to parse: selfish, not selfish.

```
Mr~Crabs is a self\makebox{}ish shellfish.
```
LaTeX Usage

Other words that need "anti-hyphenation" pre-processing are halflife, halfline, selfless, offline, offloaded, and so on.

2.4 Quotation Marks

This section explains how you typeset quotation marks. Figure 2.1 is an example from [Lamport 1994, page 13]. The word 'Convention' in this example is in single quotes and the word 'this' is in double quotes. The quotes at the start are backquotes (' and ''). The quotes at the end are the usual quotes (' and ''). Notice that output quote between 'it' and 's' is produced using a single quote in LaTeX.

To get properly nested quotations you insert a thin space where the quotes "meet." Recall that the thin space command (\,) typesets a thin space. Figure 2.2 provides a concrete example that is taken from [Lamport 1994, page 14]. Figure 2.2 provides another example. The first line of this example looks much better than the other two. Note that LaTeX parses three consecutive quotes as a pair of quotes followed by one more quote. This is demonstrated by the second line of the output, which looks terrible. The last line of the input avoids the three consecutive quotes by adding an empty group that makes explicit where the double quotes and the single quote meet. Still the resulting output doesn't look great.

Intermezzo. As a general rule, British usage prefers the use of single quotes for ordinary use. This poses a problem if an apostrophe is used for the possessive form: He said 'It is John's book.' This is why it is also acceptable to use double quotes [Trask 1997, Chapter 8].

2.5 Dashes

There are three kinds of dashes: -, –, and —. In LaTeX you get them by typing -, --, and ---. The second symbol can also be typeset with the command \textendash and the last symbol with the command \textemdash. The symbol –, which is used in mathematical expressions such as $a - b$, is not a dash. This symbol is discussed in Chapter 8. The following briefly explains how the dashes are used.

- This is the intra-word dash, which is used to hyphenate compound modifiers such as one-to-one, light-green, and so on [Trask 1997, Chapter 6]. In LaTeX you typeset this symbol as follows: -.
– This is the en-dash, which has the width of 1 en. An en is equivalent to half the *current* type size, so an en-dash is shorter in normal text than it is in large text. The en-dash is mainly used in ranges: pages 12–15 (from 12 to 15). However, the en-dash is also used to link two names that are sharing something in common: a joint Anglo–French venture [Allen 2001, page 45]. The LaTeX command \textendash and the sequence -- typeset the en-dash. When you typeset an en-dash, it looks better if you add a little space before and after. Remember that \, produces a thin space. Use this command for the horizontal space.

```
... pages~12\,--\,15 (from~12 to~15).
```
LaTeX Usage

— This is the em-dash, which has the same width as an em. An em is equal to the *current* type size. The em-dash separates strong interruptions from the rest of the sentence—like this [Trask 1997, Chapter 6]. Bringhurst [2008, page 80] prefers the en-dash to the em-dash. The LaTeX command \textemdash and the sequence --- typeset the em-dash. An em-dash at the start of a line doesn't look very good so you should tie each em-dash to the preceding word.

```
... the rest of the sentence~\textemdash
like this~\parencite[Chapter~6]{Trask:1997}.
```
LaTeX Usage

Figure 2.3 presents an example of the dashes. A few years ago I noticed that sometimes --- doesn't work with XƎTEX (even with Mapping = tex-text enabled). However, \textemdash always worked.

2.6 Full Stops

LaTeX usually treats a full stop (.) as an end-of-sentence indicator. By

```
The intra-word dash is used to hyphenate
 compound modifiers such as light-green,
 X-ray, or one-to-one. ...
The en-dash is used in ranges: pages~12--15.
The em-dash is used to separate strong
 interruptions from the rest of the
 sentence~--- like this%
 ~\cite[Chapter~6]{Trask:1997}. ...
```

Figure 2.3
Dashes

default, LaTeX inserts a bit more space after the full stop at the end of a sentence than it does between words. It also does this after other punctuation symbols. The \frenchspacing command turns this feature off. The command \nonfrenchspacing turns the feature on again. When a full stop is not the end of a sentence you need to help LaTeX a bit by inserting the space command (\␣) after the full stop.

```
Meet me at 6~p.m.\␣at the Grand Parade.
```
LaTeX Usage

However, when an uppercase letter is followed by a full stop, then LaTeX assumes the full stop is for abbreviation. For example:

```
Donald~E. Knuth developed the {\TeX} system.
```
LaTeX Usage

This convention causes a problem if an uppercase letter really is the end of a sentence. Insert a \@ before the full stop if this happens.

```
In Frank Herbert's \emph{Dune} saga,
 the Mother School of the Bene Gesserit
 is situated on the planet Wallach IX\@.
```
LaTeX Usage

LaTeX inherits its habit of putting some extra space after full stops and other punctuation symbols from TeX. Bringhurst [2008, pages 28–30] points out that there really is no reason to add such extra space for modern works. Following Bringhurst's advice, this document was typeset with \frenchspacing enabled.

2.7 Ellipsis

The command \ldots produces an ellipsis (...), which is used to indicate an omission. If the ellipsis occurs at the end of a sentence, then you still need to add an end-of-sentence marking full stop. If this happens then Felici [2012, Figure 13.15] recommends that you put the ellipsis close to the preceding text and then add the full stop.

```
Many stories start with
 'Once upon a time\ldots.'
They usually end with
 '\ldots\␣and they all lived
  happily ever after.'
```

Many stories start with 'Once upon a time....' They usually end with '... and they all lived happily ever after.'

Figure 2.4
Good borderline punctuation

```
Robert Bringhurst, author of
\emph{Elements of
      Typographic Style,}
recommends setting such
punctuation symbols in
the brighter type.
\textbf{Do as he}, or
risk getting ugly type.
```

Robert Bringhurst, author of *Elements of Typographic Style*, recommends setting such punctuation symbols in the brighter type. **Do as he**, or risk getting ugly type.

Figure 2.5
Poor borderline punctuation

```
Robert Bringhurst, author of
\emph{Elements of
      Typographic Style},
recommends setting such
punctuation symbols in
the brighter type.
\textbf{Do as he,} or
risk getting ugly type.
```

Robert Bringhurst, author of *Elements of Typographic Style*, recommends setting such punctuation symbols in the brighter type. **Do as he,** or risk getting ugly type.

2.8 Emphasis

Emphasis is a typographic tool for typesetting text in a different typeface. The idea is that this makes the text stand out. Emphasis is especially useful when introducing a new concept, such as in this paragraph.

In some documents, emphasis is implemented by typesetting text in a bold face typeface, by typesetting it in uppercase typeface, or (worse) by underlining the text. LaTeX emphasises text in paragraphs by italicising the text. Trask [1997, page 82] calls this the preferred style for emphasis. The LaTeX command for emphasis is \emph.

```
Emphasised \emph{example}.     Emphasised example.
```

2.9 Borderline Punctuation

Bold text looks darker than normal, upright text and italicised text look brighter than normal, upright text. When small punctuation symbols get caught between darker and brighter type it is time to pay attention. Robert Bringhurst, author of *Elements of Typographic Style*, recommends setting such punctuation symbols in the brighter type [Bringhurst 2008]. **Do as he**, or risk getting ugly type. Figures 2.4 and 2.5 demonstrate what you get if you follow Bringhurst's advice and what if you don't. The figures do not excel in terms of maintainability because they hardcode the author's name and the title of the work.

2.10 Footnotes and Marginal Notes

It is generally accepted that using footnotes and marginal notes should be used sparingly because they are disruptive. However, proper use of

```
Footnotes\footnote{A footnote is a note
 of reference, explanation, or comment that is
 usually placed below the text on a printed page.}
 can be a nuisance. This is especially true if
 there are many.\footnote{Like here.} The more you see
 them, the more annoying they get.\footnote{Got it?}
```

Footnotes[a] can be a nuisance. This is especially true if there are many.[b] The more you see them, the more annoying they get.[c]

[a]A footnote is a note of reference, explanation, or comment that is usually placed below the text on a printed page.
[b]Like here.
[c]Got it?

Figure 2.6
Using footnotes

marginal notes in documents with wide margins can be very effective.

Not surprisingly, LaTeX provides a command for footnotes and a command for marginal notes. Figure 2.6 demonstrates how to specify footnotes in LaTeX. A *marginal note* or *marginal paragraph* is like a footnote, but placed in the margin as on this page. The command \marginpar{⟨text⟩} puts ⟨text⟩ in the margin as a marginal note. By passing an optional argument to the command you can put different text on odd (recto/front/right) pages and on even (verso/back/left) pages. The optional argument is used for even pages and the required argument is used for odd pages. If you're using both the optional and required argument then it is easy to remember which is which: the optional argument is to the left of the required argument so it's for the left page; the required argument is for the right page. Note that narrow marginal notes may look better with ragged text, which is text that is aligned to one side only. On the right (left) pages you use ragged right (left) text. Section 2.19.2 explains how to typeset ragged text.

Avoid marginal notes in very narrow margins.

2.11 Displayed Quotations and Verses

The quote and quotation environments are for typesetting displayed quotations. The former is for short quotations; the latter is for longer quotations. Figure 2.7 shows how you use the quote environment. The command \\ in Figure 2.7 forces a line break.

The verse environment typesets poetry and verse. Figure 2.8 shows how you use the environment. In this example, the command \qquad inserts two quads. Here a quad is an amount of space that is equivalent to the *current* type size. So if you use a 12 pt typeface then a quad results in a 12 pt space in normal text. The command \\ inside the verse environment determines the line breaks. Remember that the command \, before the letter *S* inserts a thin space.

2.12 Line Breaks

In the previous section, the command \\ inserted a line break in displayed quotations and verses. The command also works inside

Figure 2.7
The quote environment

```
Blah blah blah blah blah blah blah blah blah blah blah.
\begin{quote}
    Next to the originator of a good sentence
    is the first quoter of it. \\
    \emph{Ralph Waldo Emerson}
\end{quote}
Blah blah blah blah blah blah blah blah blah blah blah.
```

Blah blah blah blah blah blah blah blah blah blah blah.

> Next to the originator of a good sentence is the first quoter of it.
> *Ralph Waldo Emerson*

Blah blah blah blah blah blah blah blah blah blah blah.

Figure 2.8
The verse environment

```
The following anti-limerick is
 attributed to W.\,S. Gilbert.
\begin{verse}
  There was an old man of St.~Bees,        \\
  Who was stung in the arm by a wasp;      \\
    \qquad When they asked, ''Does it hurt?''  \\
    \qquad He replied, ''No, it does n't,      \\
  But I thought all the while 't was a Hornet.''
\end{verse}
```

The following anti-limerick is attributed to W. S. Gilbert.

> There was an old man of St. Bees,
> Who was stung in the arm by a wasp;
>   When they asked, "Does it hurt?"
>   He replied, "No, it does n't,
> But I thought all the while 't was a Hornet."

paragraphs. An optional argument determines the extra vertical space of the line break: \\[⟨extra vertical space⟩]. A line break at the end of a page may trigger a page break. If you don't want a page break then you should use the command *. It is identical to \\ but it inhibits page breaks.

2.13 Controlling the Size

With the proper class and packages there is usually no need to change the type size of your text. However, sometimes it has its merits, e.g., when you're designing your own titlepage or environment. Table 2.5 lists the declarations and environment that change the type size. The preferred "size" for long-ish algorithms and program listings is \scriptsize. If you're using a package to typeset listings then the package usually chooses the right size. If not, it probably lets you specify the type size. Figure 2.9 shows how you change the size of text.

Declaration	Environment	Example
\tiny	tiny	Example
\scriptsize	scriptsize	Example
\footnotesize	footnotesize	Example
\small	small	Example
\normalsize	normalsize	Example
\large	large	Example
\Large	Large	Example
\LARGE	LARGE	Example
\huge	huge	Example
\Huge	Huge	Example

Table 2.5
Size-affecting declarations and environments

```
{\tiny Mumble.       \\
 \begin{normalsize}
      What?
 \end{normalsize}    \\
 \begin{Huge}
      Mumble!
 \end{Huge}  }
```

Mumble.
What?
Mumble!

Figure 2.9
Controlling the size

2.14 Seriffed and Sans Serif Typefaces

LaTeX has several commands that change the type style. Before studying these commands it is useful to study the difference between seriffed and sans serif typefaces and when to use them.

A *serif* is a little decoration at the end of some of the strokes of some of the letters. In a *seriffed* typeface the letters have serifs. Seriffed typefaces are sometimes called *roman* typefaces but in LaTeX roman means upright. In a *sans serif* typeface the letters lack serifs.

Most books use a seriffed typeface for the running text [Unger 2007, pp. 167–168] and the most popular typeface for the running text of books and reports is *(Monotype/Linotype) Times Roman* [Felici 2012], a seriffed typeface. Seriffed typefaces are also used for the running text of most papers, theses, and dissertations in science. Turabian [2007, pp. 374–375] recommends that you use a typeface that is designed for text and that you use a size in the range of 10–12 pt, with 12 pt being the preferred size. Admittedly, the being designed for text is a bit vague but Turabian [2007] give two examples, both of which are seriffed.

As lines get longer and longer, seriffed typefaces are easier to read and make fast reading easier [Unger 2007]. Sans serif typefaces may look better on the screen but the ultimate criterion for printed matter is how the text looks in print, so never choose the typeface for your printed text based on how it looks on the screen.

If a typeface family has a seriffed and sans serif typeface of the same type size (point size), then the seriffed typeface usually requires more horizontal space [Unger 2007]. Stated differently, sans serif typefaces are usually more efficient when it comes to saving space. This may be exploited by using sans serif typefaces in captions, in brochures, in short narrow columns, or on road signs [Unger 2007].

If you don't change the typeface then LaTeX will typeset the body of your document in *Computer Modern*. An example of *Computer Modern* may be found in Table 2.6, further on in this chapter.

2.15 Small Caps Letters

Small caps letters are used to typeset acronyms and abbreviations. Their shape is the same as uppercase letter but their height is smaller, which lets them blend in better with the rest of the text. For example, compare NO SHOUTING with NO SHOUTING. The latter is easier on the eye.

Adding extra space uniformly to the left and right of characters in a passage of text is called tracking or letterspacing. The extra space that is added per letter is called the tracking space. Tracking passages of small caps text is a common technique to improve the legibility. For example NON-SPACED SMALL CAPS is not spaced, whereas SPACED SMALL CAPS is letterspaced.

The command \textsc typesets lowercase letters in small caps. The easiest way to automatically letterspace such text is to use the microtype package with the option tracking=smallcaps. After this all small caps text will be letterspaced.

```
\textsc{No shouting}.              NO SHOUTING.
```

The microtype package also provides character protrusion (margin kerning) and font expansion. Character protrusion adjusts the characters at the margins of the text. Font expansion uses narrow or wider font versions so as to make the overall appearance of the text more uniform, avoiding long cramped, dark lines with many characters and long loose, bright lines with few characters. As a side-effect, font expansion may also be used to choose better hyphenation points [Schlicht 2010]. This document was typeset using the microtype package with the following options.

```
\usepackage[final,tracking=smallcaps,          LaTeX Usage
            expansion=alltext,protrusion=true]{microtype}
```

Bringhurst [2008, page 30] recommends that you add 5–10% of the type size (point size) for the tracking space. The microtype package expects the extra tracking in thousands of the type size. The following sets the tracking space to 5% for the sc (small caps) shape.

```
\SetTracking{encoding=*,shape=sc}{50}          LaTeX Usage
```

Most microtype users agree that the package improves the appearance of their documents.

Declaration	Command	Example
\mdseries	\textmd	Medium Series
\normalfont	\textnormal	Normal Style
\rmfamily	\textrm	Roman family
\upshape	\textup	Upright Shape
\itshape	\textit	*Italic Shape*
\slshape	\textsl	*Slanted Shape*
\bfseries	\textbf	**Boldface Series**
\scshape	\textsc	SMALL CAPS SHAPE
\sffamily	\textsf	Sans Serif Family
\ttfamily	\texttt	Typewriter Family

Table 2.6

Type style affecting declarations and commands. The last column shows the result in *Computer Modern* (LATEX's default typeface). The first four lines usually correspond to the default style. The first nine typefaces are proportional. They may have glyphs with different widths, e.g., compare *M* and *i*. Small caps letters are useful for abbreviations. The last typeface is non-proportional, which is useful in program listings.

2.16 Controlling the Type Style

Changing the type size is hardly ever needed in an article, thesis, report, or book. Changing the type style is required much more, but usually this is done automatically by the commands that typeset the title of your document, the section titles, the captions, and so on.

There are ten LATEX type style affecting declarations. Each declaration has a command that takes an argument and applies the type style of the declaration to the argument. The arguments cannot have paragraph breaks. The declarations and commands are listed in Table 2.6.

Intermezzo. If you really must change the type style of your text then it is probably for a specific purpose. For example, to change the type style of a newly defined word, to change the type style of an identifier in an algorithm, and so on. Rather than hard-coding the style in your input, it is better if you define a user-defined command that typesets your text in the required style and use the command to typeset your text. The command's name should reflect its purpose. For example \identifier to typeset an identifier in an algorithm, \package to typeset the name of a LATEX package, and so on. Using this approach improves maintainability. For example, if you want to change the type style of all identifiers in your text then you only need to make changes in the definition of the command that typesets identifiers. Defining your own commands is discussed in Chapter 11.

2.17 Abbreviations

This section is about abbreviations. It provides some guidelines about their spelling and how to typeset them in LATEX.

2.17.1 *Initialisms*

Abbreviations that are made up of the initial letters of the abbreviated words are called *initialisms*. Non-standard initialism are usually written with a full stop after each part in the abbreviation: Ph.D. (Philosophiae

Figure 2.10

Finer points of typesetting abbreviations

```
In~2010 Prof.~Donald Knuth was invited to the
annual TeX User Group Conference in San Francisco, Ca.\␣
to speak about a revolutionary successor to \TeX.
This remarkable system is entirely menu driven and
incorporates facilities for social networking.
Pronouncing the name involves making the sound of a bell.
```

Doctor), D.Phil. (Doctor of Philosophy), M.Sc. (Master of Science), and so on. However, if the initialisms are standard, then you omit the full stops, so B.B.C. becomes BBC, 4 G.L. (fourth-Generation Language) becomes 4GL, and Ph.D. becomes Ph D (in LaTeX Ph~D). Bringhurst [2008, page 48] recommends typesetting abbreviations with more than two uppercase letters in spaced small capitals: SPACED SMALL CAPS. Section 2.15 explains how to get spaced small caps.

Some authors recommend that you letterspace Uniform Resource Locators (URLS), phone numbers, and email addresses because they are not words. See for example [Bringhurst 2008] or [Hedrick 2003].

Abbreviations of personal names such as D. E. K., J. F. K., J. S. B., and the like should not be letterspaced.

2.17.2 *Acronyms*

An *acronym* is an initialism that is pronounced as a word. For example, radar (RAdio Detection And Ranging), sonar (SOund Navigation And Ranging), NASA (National Aeronautics and Space Administration), and EBCDIC (Extended Binary Coded Decimal Interchange Code); but not ACM (Association for Computing Machinery), BBC (British Broadcasting Corporation), and RSVP (Répondez S'il Vous Plaît). Note that not all acronyms are spelt with uppercase letters; if you're not certain, look up the spelling. Since acronyms are just a special form of initialisms, we should follow Bringhurst's advice, and write them with small caps if they are spelt with (two or more) uppercase letters.

2.17.3 *Shortenings*

A word that is abbreviated by taking the first few letters of that word is called a *shortening*. To avoid ambiguity, shortenings are usually written with a full stop at the end of each part. For example, p. (page), proc. (proceedings), sym. (symposium), fig. (figure), Feb. (February), Prof. (Professor), and so on. The abbreviation pp. is for pages.

Remember that LaTeX inserts a little extra white space after a full stop if \frencspacing isn't enabled. If an abbreviation is not at the end of a sentence and ends with a full stop then this extra space may look bad. To suppress the extra white space you have to hardcode a space command (\␣) after the abbreviation or tie the abbreviation and the following word. Figure 2.10 provides a small example.

2.17.4 *Introducing Abbreviations*

The first time you introduce an abbreviation you should explain it. Most authors first spell out the abbreviation and then provide the abbreviation in parenthesis. The acronym package provides some support for defining and referencing abbreviations in a consistent style. This is done using the standard label-referencing technique. The package provides commands for singular and plural versions of abbreviations and for abbreviated and unabbreviated versions.

Page 4 of this book introduces an acronym for integrated development environments. This text was generated by the following input.

```
... many \acp{IDE} ...
```
LaTeX Input

The command \acp in this example is provided by the acronym package. The command introduces the plural version of an abbreviation. The acronym package also provides the \ac command, which introduces the singular version of an abbreviation. The argument IDE of the \acp command is the label of the acronym. Some other part of the input associates the label IDE with the abbreviated version 'IDE' and the expanded version 'Integrated Development Environment.' This was (essentially) done as follows:

```
\acro{IDE}[\textsc{ide}]%
    {Integrated Development Environment}
```
LaTeX Input

When this book was generated and the command \acp was used in the second last input, this was the first time the label IDE was referenced, which is why it resulted in the following output.

```
... many Integrated Development Environments (IDEs) ...
```
LaTeX Output

The label IDE is also referenced in other locations in the input, but when that happens it always results in the abbreviated version of the acronym: IDE. More information about the acronym package may be found in the package documentation [Oetiker 2010].

2.17.5 *British and American Spelling*

There are differences between American and British usage in time abbreviations. According to Trask [1997] Americans write 10:05 AM (Ante Meridiem) for five past ten in the morning and 13:15 PM (Post Meridiem) for a quarter past one in the afternoon. British spelling prefers 10.05 a.m. and 13.15 p.m. [Trask 1997]. Felici [2012] notices that Americans have also started using the British form.

For titles such as Mister, Doctor, and so on, British and American usage differ. Britsh usage is the same as for shortenings. For example, Mr Happy, Dr Who, and Fr Dougal McGuire. Americans add the full stop: Mr. Ed, Dr. Quinn, Medicine Woman, and Fr. Bob Maguire.

Table 2.7
Latin abbreviations. The first column lists the abbreviations, the second the original Latin meaning, and the last the English translation. Note that the abbreviations at the bottom of the table are slanted. This is intensional and preferred usage.

Abbreviation	Latin meaning	English meaning
e.g.	exempli gratia	for example
i.e.	id est	that is/in other words
etc.	et cetera	and so forth
viz.	videlicet	that is to say/namely
cf.	confer	compare
et al.	et aluis	and others

2.17.6 *Latin Abbreviations*

This section studies some Latin abbreviations that are commonly used in scientific writing. Table 2.7 presents the more commonly occurring abbreviations, their Latin meaning, and the English translation.

Note that some abbreviations are typeset in italics. This is not by accident: this is how they should be typeset—but conventions may differ from field to field. Also note that the *al* in *et al.* gets a full stop because it is an abbreviation of *aluis* but that the *et* does not get a full stop because it is already spelt out in full. Remember Bringhurst's advice and put the full stop inside the argument of \emph: \emph{et al.} Finally note that etc. is short for et cetera, not for ectcetra.

Trask [1997] discourages these abbreviations. Trask continues by pointing out that writing statements like the following are wrong because the reader should be invited to consult the reference.

> The Australian language Dyirbal has a remarkable gender system, *cf.* [Dixon 1972].

Trask proposes the following solution.

> The Australian language Dyirbal has a remarkable gender system; see [Dixon 1972].

Abbreviations such as etc., i.e., and e.g. require additional punctuation [Strunk, and White 2000]:

○ Abbreviations such as BBC, NBC, etc., are called initialisms.
○ Shortenings, i.e., abbreviations that are formed by taking the first letters of the abbreviated word, usually end with a full stop.
○ Abbreviations are not always spelt te same, e.g., Ph.D. and Ph D.

2.17.7 *Units*

The Système International d'Unités/International System of Units (SI) provides rules for consistent typesetting of quantities of units. Heldoorn [2007] provides a summary of these rules. The following is a summary of the main rules.

○ The base unit symbols are printed in upright roman: g (gram), m (metre), t (tonne), ... Exceptions are unit symbols that are spelt in Greek and the symbols for inch, degrees, seconds, and so on.

| ```
Fill in the missing word.\\
Fill in the missing
  \phantom{word}.
``` | Fill in the missing word.<br>Fill in the missing      . | **Figure 2.11**<br>The \phantom command |

○ The first letter of the unit symbol is uppercase if it is derived from a proper name: Å (Ångström), N (Newton), Pa (Pascal), ...

○ The plural form of the base unit symbol is the same as the singular.

○ The base unit symbols do not receive an end-of-abbreviation full stop.

Needless to say, it is important that you typeset quantities of units correctly and consistently. The hard way is doing it by hand. The easy way is doing it with LATEX.

At the moment of writing the most popular package for specifying SI units is the siunitx package [Wright 2011].

○ It provides support to configure how the SI units are typeset. For example, kg m s^{-1}, versus kg m s^{-1}, versus kg m/s, and so on.

○ It provides commands to typeset quantities of units: \SI[mode=text] {1.23}{\kilogram} will give you 1.23 kg and \SI{1.01}{\kilogram} will typeset 1.01 kg in the default typesetting mode.

○ The package provides macros to typeset lists of quantities in a given unit. For example \SIlist{0.1;0.2;1.0}{\milli\metre} gives you 0.1, 0.2 and 1.0 mm if the default typesetting mode is text. If you add the option list-final-separator={, and~} then you get 0.1, 0.2, and 1.0 mm.

○ By default, unit symbols are typeset using the default math roman font but you can also use different fonts.

Discussing the entire siunitx package is beyond the scope of this book. The interested reader is referred to the package documentation [Wright 2011] for further information.

2.18 Phantom Text

Some commands don't typeset anything with ink but do affect the horizontal and vertical spacing. The following is the first of three useful versions.

This command "typesets" its argument using invisible ink. The dimensions of the box are the same as the dimensions required for typesetting ⟨stuff⟩.   ☑

Figure 2.11 demonstrates how you use the command. The \hphantom and \vphantom commands are horizontal and vertical versions of the \phantom command. The following explains how they work.

\hphantom{⟨stuff⟩}

This is the horizontal version of the \phantom command. The command creates a box with zero height and the same width as its argument, ⟨stuff⟩.   ☑
```

Figure 2.12
The center environment

```
\begin{center}
  Blah.\\
  Blah blah blah.

  Blah blah blah blah blah
  blah blah blah blah blah
  blah blah blah blah blah.
\end{center}
```

Blah.
Blah blah blah.
Blah blah blah blah blah blah
blah blah blah blah blah blah
blah blah blah.

\vphantom{⟨stuff⟩}

This is the vertical version of the \phantom command. The command creates a box with zero width and the same height as its argument, ⟨stuff⟩. It is especially useful for getting the right size for delimiters such as parentheses in mathematical formulae that span multiple lines. This is explained in more detail in Section 8.8.1. ☑

2.19 Alignment

This section studies three commands and two environments that change the text alignment. The first command centres text. The second and third command align text to the left and to the right. The first of the environments is the tabular environment, which typesets row-based content with horizontal alignment positions (columns). The last environment is the tabbing environment. This environment lets you define horizontal alignment (tab) positions and lets you position text relative to these alignment positions.

2.19.1 *Centred Text*

The center environment centres text. The example in Figure 2.12 demonstrates the environment. The example is inspired by Iggy Pop.

2.19.2 *Flushed/Ragged Text*

The flushleft environment and the \raggedright declaration typeset text that is aligned to the left. Likewise, the flushright environment and \raggedleft declaration typeset text that is aligned to the right. The example in Figure 2.13 shows the effect of the flushleft environment.

2.19.3 *Basic tabular Constructs*

The tabular environment typesets text with rows and alignment positions for columns. The environment also has siblings called tabular* and array. The tabular* environment works similar to tabular but it takes an additional argument that determines the width of the resulting construct. This environment is explained in Section 2.19.5. The

```
\begin{flushleft}
  Blah.\\
  Blah blah blah.

  Blah blah blah blah blah
  blah blah blah blah blah
  blah blah blah blah blah.
\end{flushleft}
```

Blah.
Blah blah blah.
Blah blah blah blah blah blah
blah blah blah blah blah blah
blah blah blah.

Figure 2.13
The flushleft environment

array environment can only be used in math mode. The tabular and tabular* environments can be used in both text and math mode.

The remainder of this section introduces the tabular environment. This introduction should more than likely suffice for day-to-day usage. A more detailed presentation is provided in Section 2.19.5.

In its simplest form the tabular environment is used as follows.

```
\begin{tabular}[⟨global alignment⟩]
            {⟨column alignment⟩}
   ⟨text⟩ & ⟨text⟩ & ... & ⟨text⟩ \\
   ...
   ⟨text⟩ & ⟨text⟩ & ... & ⟨text⟩ \\
   ⟨text⟩ & ⟨text⟩ & ... & ⟨text⟩
\end{tabular}
```
LaTeX Usage

The body of the environment contains a sequence of rows that are delimited by linebreaks (\\). Each row is a sequence of alignments tab-delimited ⟨text⟩. The i-th ⟨text⟩ in a row corresponds to the i-th column. The following explains the arguments of the environment:

⟨**global alignment**⟩
This optional argument determines the vertical alignment of the environment. Allowed values are t (align on the top row), c (align on the centre), or b (align on the bottom row). The default value of this argument is c. ☑

⟨**column alignment**⟩
This argument determines the column alignment and additional decorations. For day-to-day usage, the following options are relevant.

   l   This option corresponds to a left-aligned column.

   r   This corresponds to a right-aligned column.

   c   This corresponds to a centred column.

   p{⟨width⟩} This option corresponds to a top-aligned ⟨width⟩-wide column that is typeset as a paragraph in the "usual" way. Some commands such as \\ are not allowed at the top level.

   | This option does not correspond to an actual column but results in additional decoration. It results in a vertical line drawn at at the "current" position. For example, if ⟨column alignment⟩ is l|cr then there will be a vertical line separating the first two columns. Using this option is discouraged because the vertical lines usually distract. ☑

The tabular environment also defines the following commands,

```
\begin{tabular}{l|crp{3.1cm}}
    \hline
       1 &   2 &   3
         & Box me in,
           but not too
           tight, please.
    \\\hline
       11 &  12 &  13 & Excellent.
    \\ 111 & 112 & 113 & Thank you!
    \\\hline
\end{tabular}
```

1	2	3	Box me in, but not too tight, please.
11	12	13	Excellent.
111	112	113	Thank you!

which may be used inside the environment. You can only use these commands at the start of a row.

\hline

This command inserts a horizontal rule. The command may only be used at the *start* of a row. ☑

\cline{⟨number₁⟩}{⟨number₂⟩}

This draws a horizontal line from the start of column ⟨number₁⟩ to the end of column ⟨number₂⟩. ☑

\vline

This results in a vertical line. The command may only be used if the column is aligned to the left, to the right, or to the centre. ☑

Figure 2.14 presents a simple example of the tabular environment. The example shows all alignments and the paragraph feature.

Note that line breaks are inserted automatically inside p-type columns. Line breaks are not allowed in columns aligned with l, r, or c.

Intermezzo. The column alignment option | and the commands \hline, \cline, and \vline are irresistible to new users. This may be because most examples of the tabular environment involve the option and these commands. It is understandable that new users want to repeat this, especially when they're not aware that using the option and the commands in moderation is better because the grid lines are dazzling and distracting. Chapter 6 provides some guidelines on how to design good tables.

Regular $m \times n$ tables with the same alignment in the same column are rare. The following command lets you join columns within a row and override the default alignment.

\multicolumn{⟨number⟩}{⟨column alignment⟩}{⟨text⟩}

This inserts ⟨text⟩ into a *single* column that is formed by combining the next ⟨number⟩ columns in the current row. The alignment of the column is determined by ⟨column alignment⟩. This command is especially useful for overriding the default alignment in column headings of a table. An example is presented in the next section. ☑

2.19.4 *The booktabs Package*

The booktabs package adds some extra functionality to the tabular environment. The package discourages vertical grid lines. Using the booktabs package results in better looking tables.

- The package provides different commands for different rules.
- The package provides different rules that may have different widths.
- The package provides commands for temporarily/permanently changing the width.
- The package has a command that adds extra line space.
- The package is compatible-ish with the colortbl package, which is used to specify coloured tables.

The booktabs package provides the following commands. The first four commands take an option that specifies the width of the rule. The first four commands can only be used at the start of a row.

\toprule[⟨width⟩]

This typesets the full horizontal rule at the top the table.   ☑

\bottomrule[⟨width⟩]

This typesets the full horizontal rule at the bottom of the table.   ☑

\midrule[⟨width⟩]

This typesets the remaining full horizontal rules in the table.   ☑

\cmidrule[⟨width⟩]{⟨number₁⟩-⟨number₂⟩}

This typesets a partial horizontal rule. The rule is supposed to be used in the middle of the table. It ranges from the start of column ⟨number₁⟩ to the end of column ⟨number₂⟩.   ☑

\addlinespace[⟨height⟩]

This command is usually used immediately after a line break and it inserts more vertical line space to the height of ⟨height⟩.   ☑

Figure 2.15 demonstrates how to use the booktabs-provided rule commands. The resulting output is presented in Figure 2.16. Notice that the inter-linespacing is much better than the output in Figure 2.14. Also notice the different widths of the rules.

2.19.5 *Advanced tabular Constructs*

Using basic tabular constructs usually suffices for day-to-day typesetting. This section explains the techniques that give you the power to typeset more advanced tabular constructs.

The following starts by presenting two addition column options. This is followed by some style parameters that control the default size and spacing of the tabular, tabular*, and array environments. The column options are as follows.

*{⟨number⟩}{⟨column options⟩}

This inserts ⟨number⟩ copies of ⟨column options⟩. For example, *{2}{ lr} is equivalent to lrlr.   ☑

@{⟨text⟩}

This is called an @-expression. It inserts ⟨text⟩ at the current position.

```
\begin{tabular}[c]{lrrp{47mm}}
  \toprule \multicolumn{1}{r}{\textbf{Destination}}
        & \multicolumn{1}{r}{\textbf{Duration}}
        & \multicolumn{1}{r}{\textbf{Price}}
        & \multicolumn{1}{r}{\textbf{Description}}
\\\midrule
  Cork City
  &  7 Days & \euro 300.00
  & Visit Langer Land. Price includes visits
    to Rory Gallagher Place and de Maarkit.
\\ Dingle
  &  8 Days & \euro 400.00
  & Have fun with Fungie.
\\\bottomrule
\end{tabular}
```

Figure 2.16

Output of booktabs package. The input of this figure is listed in Figure 2.15. Clearly, booktabs rules rule.

Destination	Duration	Price	Description
Cork City	7 Days	€300.00	Visit Langer Land. Price includes visits to Rory Gallagher Place and de Maarkit.
Dingle	8 Days	€400.00	Have fun with Fungie.

This is useful if you want to add certain text or symbols at the given position. For example @{.} inserts a full stop at the current position.

LaTeX normally inserts some horizontal space before the first column and after the last column. It inserts twice that amount of space between adjacent columns. However, this space is suppressed if an @-expression precedes or follows a column option. For example, if ⟨column alignment⟩ is equal to @{}ll@{}l@{} then this suppresses the horizontal space before the first column, after the last column, and between the second and last column. The length \tabcolsep controls the extra horzontal space that is inserted. The value of the command is half the width that is inserted between columns.

A horizontal spacing command in an @-expression controls the separation of two adjacent columns. For example, @{\hspace{⟨width⟩}} inserts a horizontal ⟨width⟩-wide space.

Finally, @-expressions may also adjust the default column separation. The \extracolsep{⟨width⟩} adds additional horizontal ⟨width⟩-wide space between subsequent columns. *However*, additional width is never inserted before the first column. The \extracolsep{\fill} inserts the maximum possible amount of horizontal space. This is useful if you want to extend the width to the maximum possible width.

The columns in the second table in Figure 2.17 are spread out evenly with an @-expression. The third table adds the usual space to the start of the first and the end of the last column. The first table is added for comparison. ☑

```
\begin{tabular*}{3cm}{@{}lcr@{}}
  \toprule M & M & M \\\bottomrule
\end{tabular*}
\begin{tabular*}{3cm}
      {@{\extracolsep{\fill}}%
       lcr%
       @{\hspace{0pt}}}
  \toprule M & M & M \\\bottomrule
\end{tabular*}
\begin{tabular*}{3cm}
      {@{\hspace{\tabcolsep}}%
       @{\extracolsep{\fill}}%
       lcr%
       @{\hspace{\tabcolsep}}}
  \toprule M & M & M \\\bottomrule
\end{tabular*}
```

M M M	M M M	M M M

Figure 2.17
Controlling column widths with an @-expression. The output is spaced out for clarity.

The following commands control the default appearance of tabular, tabular*, and array environments.

\arraycolsep

The value of this length command is equal to half the default horizontal distance between adjacent columns in the array environment. This amount of space is also equal to the default horizontal space inserted before the first column and after the last column. ☑

\tabcolsep

The value of this length command is equal to half the default horizontal distance between adjacent columns in the tabular and tabular* environments. Again, this is equal to the default horizontal space that is inserted before the first column and after the last column. ☑

\arrayrulewidth

The value of this length command is the width of the lines resulting from a | in the ⟨column options⟩ argument and the lines resulting from the commands \cline, \hline, and \vline. ☑

\doublerulesep

The value of this length command is the distance between two adjacent lines resulting from a || in the ⟨column options⟩ argument or two adjacent lines resulting from the \hline command. ☑

\arraystretch

This command determines the distance between successive rows. It defaults to 1 and "multiplying" it by x results in rows that are x times further apart. So, by redefining this command to 0.50 you halve the row distance. Redefining commands is explained in Chapter 11. ☑

2.19.6 The tabbing Environment

The tabbing environment is useful for positioning material relative to user-defined alignment positions. The remainder of this section

Figure 2.18

The tabbing environment

```
\begin{tabbing}
From \=here to \=there \\
    \>and \>then\\
    \>\>all\\
    \>the \>way\\
back \>to \>here.
\end{tabbing}
```

```
From here to there
            and     then
                    all
            the     way
back  to    here.
```

Figure 2.19

Advanced tabbing

```
\begin{tt}\begin{tabbing}
AAA\=AAA\=AAA\=AAA \kill
FUNC euc( INT a,
          INT b ): INT \\
BEGIN \+ \\
  WHILE (b != 0) DO \\
  BEGIN \+ \\
    INT rem = a MOD b;\\
    a = b; \\
    b = rem; \- \\
  END \\
  RETURN a; \- \\
END;
\end{tabbing}\end{tt}
```

```
FUNC euc( INT a, INT b ): INT
BEGIN
   WHILE (b != 0) DO
   BEGIN
      INT rem = a MOD b;
      a = b;
      b = rem;
   END
   RETURN a;
END;
```

describes some basic usage of the environment. The reader is referred to [Lamport 1994, pages 201–203] for more detailed information.

The tabbing environment can only be used in *paragraph mode* (the "usual mode"). It produces lines of text with alignment in columns based upon *tab positions*.

`\=`

Defines the next tab (alignment) position. ☑

`\\`

Inserts a line break and resets the next tab position to the value of `left_margin_tab`. ☑

`\kill`

Throws away the current line but remembers the tab positions defined with `\=`. ☑

`\+`

Increments `left_margin_tab`. ☑

`\-`

Decrements `left_margin_tab`. ☑

`\>`

Move to the next tab stop. ☑

Figures 2.18 and 2.19 present two examples of the tabbing environment. The examples do not demonstrate the full functionality of the environment.

```
\usepackage[dutch,british]{babel}
⋮
\selectlanguage{dutch}
% Dutch text here.
 Nederlandse tekst hier.

\selectlanguage{british}
% Engelse tekst hier.
 English text here.
```

Figure 2.20
Using the babel package

2.20 Language Related Issues

As suggested by its title, this section is concerned with language related issues. The remaining three sections deal with hyphenation, foreign languages, and spelling.

2.20.1 *Hyphenation*

LaTeX's (TeX's really) automatic hyphenation is second to none. However, sometimes even TeX gets it wrong. There are two ways to overcome such problems.

○ The command \- in a word tells LaTeX that it may hyphenate the word at that position.

```
Er\-go\-no\-mic has three hyphenation positions.        LaTeX Usage
```

○ Specifying the same hyphenation patterns is messy and prone to errors. Using the \hyphenation command is a much cleaner solution. This command takes one argument, which should be a comma-separated list of words. For each word you can put a hyphen at the (only) possible, desired, or allowed hyphenation positions. You may use the command several times. The following is an example.

```
\hyphenation{fortran,er-go-no-mic}        LaTeX Usage
```

2.20.2 *Foreign Languages*

The babel package supports multi-lingual documents. The package supports proper hyphenation, switches between different languages in one single document, definition of foreign languages, commands that recognise the "current" language, and so on. Figure 2.20 provides a minimal example. Rik Kabel kindly informed that X∃TEX users use the polyglossia package instead of babel. One of the advantages of the polyglossia package is that it automatically loads the bidi package when bi-directional scripts are used.

2.20.3 *Spell-Checking*

LaTeX does not support automatic spell-checking. Note that spell-checking isn't trivial anyway because commands may generate text. Text may come from external files, so make sure you spell-check your bibliography files.

However, most modern IDES have a spell checker. The ispell program, which can be run from the command line, has a LaTeX spell-check mode. The -t flag tells the command that the input is LaTeX.

```
$ ispell -l -t -S input.tex | sort -u                    Unix Session
```

CHAPTER 3

Lists

THIS CHAPTER is about *lists*. Here, a *list* is a sequence of labelled *items*. LaTeX has three built in environments supporting *unordered* lists, *ordered* lists, and *description* lists.

LaTeX marks each item in the output list by giving it a *label* that precedes the item. The items are typeset into paragraphs with hanging indentation, which means the paragraphs are indented a bit further than the surrounding text.

In unordered lists, all labels are the same and the item order is irrelevant. Usually the labels are bullet points, asterisks, dashes, ... Most people refer to such lists as bullet points.

In an ordered list the order of the items *does* matter. Each label indicates the order of its item in the list. You could say that the items are numbered by the labels. Usually, the labels are arabic numbers (1, 2, 3, ...), lowercase roman numerals (i, ii, iii, ...), lowercase letters (a, b, ..., z), and so on.

In a description list, the order of the items may or may not matter. In such lists, each label describes its item.

3.1 Unordered Lists

The itemize environment creates an unordered list. In the body of the environment you start each item in the list using the \item command. Each itemize environment should have at least one \item. Nested itemised lists are possible, but the nesting level is limited. Figure 3.1 shows how you use the itemize environment.

Each item in the output list is preceded by its label. Usually, the shape of the top-level label is a bullet point but the shape may depend on your document class and your packages.

The command \labelitemi determines the shape of the label of the top-level items. Likewise, \labelitemii is for labels of subitems, \labelitemiii for the labels of subsubitems, and \labelitemiv for the labels of subsubsubitems. By redefining these commands, you may change the appearance of the labels. You may redefine an existing command with the \renewcommand* command. The following sets the shape of the labels at the top level to a plus sign and the labels at level four to a minus sign. The command \renewcommand* is discussed in more detail in Chapter 11.

Figure 3.1

The itemize environment. Notice that the labels of the nested list are different from the labels of the top-level list.

```
\begin{itemize}
\item First item.
\item Second item.
      Text works as usual here.
\item Third item is a list.
      Different labels here.
      \begin{itemize}
      \item First nested item.
      \item Second item.
      \end{itemize}
\end{itemize}
```

- First item.

- Second item. Text works as usual here.

- Third item is a list. Different labels here.

  – First nested item.
  – Second item.

Figure 3.2

Changing the item label. The command \labelitemi defines the label for top-level itemised lists. By default it gives a bullet. By redefining the command we get a plus in the first list. The new definition is local to the first environment. Therefore, the second list still has a bullet-shaped label.

```
\begin{itemize}
\renewcommand*\labelitemi{+}
\item Label is plus.
\end{itemize}

\begin{itemize}
\item Default label.
\end{itemize}
```

+ Label is plus.

- Default label.

```
                                                  LaTeX Usage
\renewcommand*\labelitemi{+}
\renewcommand*\labelitemiv{-}
```

If you only want to change the appearance of a given label for a single list then the easiest way is to redefine the \labelitemi command at the start of the body of the list environment. Since the body of the environment is automatically put in a group, this keeps the new definition of \labelitemi local to the environment.[1] Figure 3.2 shows how to change the appearance of the label at Level 1 and keep the change local to the environment.

3.2 Ordered Lists

The enumerate environment creates an ordered list. It works just as the itemize environment but this time the labels are numbers, letters, or roman numerals, and such. Figure 3.3 demonstrates the environment.

As with the appearance of the labels in the itemize environment you may also change the appearance of the labels in the enumerate environment. This time the appearance depends on the four labelling commands \labelenumi, \labelenumii, \labelenumiii, and \labelenumiv. Each of these commands depends on a *counter* that counts the items at its level. Counters are explained in further detail in Chapter 12. The top level items are counted with the counter enumi, the second level items with enumii, the third level items with enumiii,

---

[1] See Section 2.1.2 for further information about the merits of groups.

```
\begin{enumerate}
\item First item.
\item Second item.
\item Third item is a list.
      \begin{enumerate}
      \item First nested item.
      \item Second item.
      \end{enumerate}
\end{enumerate}
```

1. First item.

2. Second item.

3. Third item is a list.

    (a) First nested item.

    (b) Second item.

Figure 3.3

The enumerate environment. Notice that the labels of the top-level list are numeric, whereas the labels of the nested list are parenthesised lowercase letters.

and the fourth level items with `enumiv`. These counters are managed by the labelling commands. When a labelling command typesets its label, it uses the corresponding counter and typesets the label in a certain style. The style is hard-coded in the command. Typical styles are (1) arabic numbers, (2) lowercase or uppercase roman numerals, and (3) lowercase or uppercase letters. Implementing the typesetting of the label in these styles is done with the commands `\arabic`, `\roman`, `\Roman`, `\alph`, and `\Alph`, which typeset their counter using arabic numbers, lowercase roman numerals, uppercase roman numerals, lowercase letters, and uppercase letters. The following demonstrates how you get lowercase roman numerals for the labels at the top level and numbers for the labels at the third level.

```
\renewcommand*\labelenumi{\roman{enumi}}
\renewcommand*\labelenumiii{\arabic{enumiii}}
```
LaTeX Usage

3.3 The enumerate Package

Redefining labelling commands is tedious and prone to error. The `enumerate` package provides a high-level interface to LaTeX's default mechanism for selecting the labels of enumerated lists. Basically, the package redefines the `enumerate` environment. The resulting environment has an optional argument that determines the style of the labels of the lists. For example, using the option A results in labels that are typeset using the command `\Alph`. Likewise the options a, I, i, and 1 result in labels that are typeset using the commands `\alph`, `\Roman`, `\roman`, and `\arabic`.

However, the package is more flexible and also allows different kinds of labels. Figure 3.4 provides an example. The interested reader is referred to the package documentation [Carlisle 1999a] for further details.

3.4 Description Lists

The `description` environment creates a labelled list. The labels are passed as optional arguments to the `\item` command. Figure 3.5 provides an example of how to use the `description` environment.

Figure 3.4

Using the enumerate package. The enumerated list is created with the environment enumerate, which is redefined by the enumerate package. The optional argument defines labels of the form Item-⟨uppercase letter⟩. The labels are typeset relative to the start of the hanging paragraphs. For long labels, such as in this example, this makes them protrude into the margin of the surrounding text.

```
\usepackage{enumerate}
\begin{document}
...
Surrounding text here.
\begin{enumerate}
        [\textbf{{Item}-A}]
\item The first of two
      hanging paragraphs.
\item The second of two
      hanging paragraphs.
\end{enumerate}
```

Surrounding text here.

Item-A The first of two hanging paragraphs.

Item-B The second of two hanging paragraphs.

Figure 3.5

Using the description environment. The environment works almost the same as the itemize and enumerate environments. The only difference is that this time you provide the labels for the list as optional arguments of the \item command.

```
Kurosawa films include:
\begin{description}
\item[Kagemusha]
    When a powerful warlord
    in medieval Japan dies,
    a poor thief is recruited
    to impersonate him. ...
\item[Yojimbo]
    A crafty ronin comes
    to a town divided by
    two criminal gangs. ...
\item[Sanshiro Sugata]
    A young man struggles to
    learn the nuance and
    meaning of judo. ...
\end{description}
```

Kurosawa films include:

Kagemusha When a powerful warlord in medieval Japan dies, a poor thief is recruited to impersonate him. ...

Yojimbo A crafty ronin comes to a town divided by two criminal gangs. ...

Sanshiro Sugata A young man, struggles to learn the nuance and meaning of judo. ...

3.5 Making your Own Lists

LaTeX's list environment lets you define your own lists.

`\begin{list}{⟨label commands⟩}{⟨formatting commands⟩} ⟨item list⟩ \end{list}`

Here ⟨label formatting commands⟩ typesets the labels. For ordered lists you may need to define a dedicated counter that keeps track of the numbers of the labels. How to do this is explained later in this chapter. The commands in ⟨formatting commands⟩ format the resulting list. The formatting depends on *length* commands. These commands determine lengths and widths that are used to construct the resulting list. For example, the distance between adjacent items, the distance between a label and its item, and so on. Figure 3.6 depicts the relevant length commands and how they determine the formatting of the list. The picture is based on [Lamport 1994, Figure 6.3]. The horizontal length commands are rigid lengths. As the name suggests the resulting dimensions are fixed. The vertical length commands are rubber lengths. Rubber lengths result in dimensions that shrink or

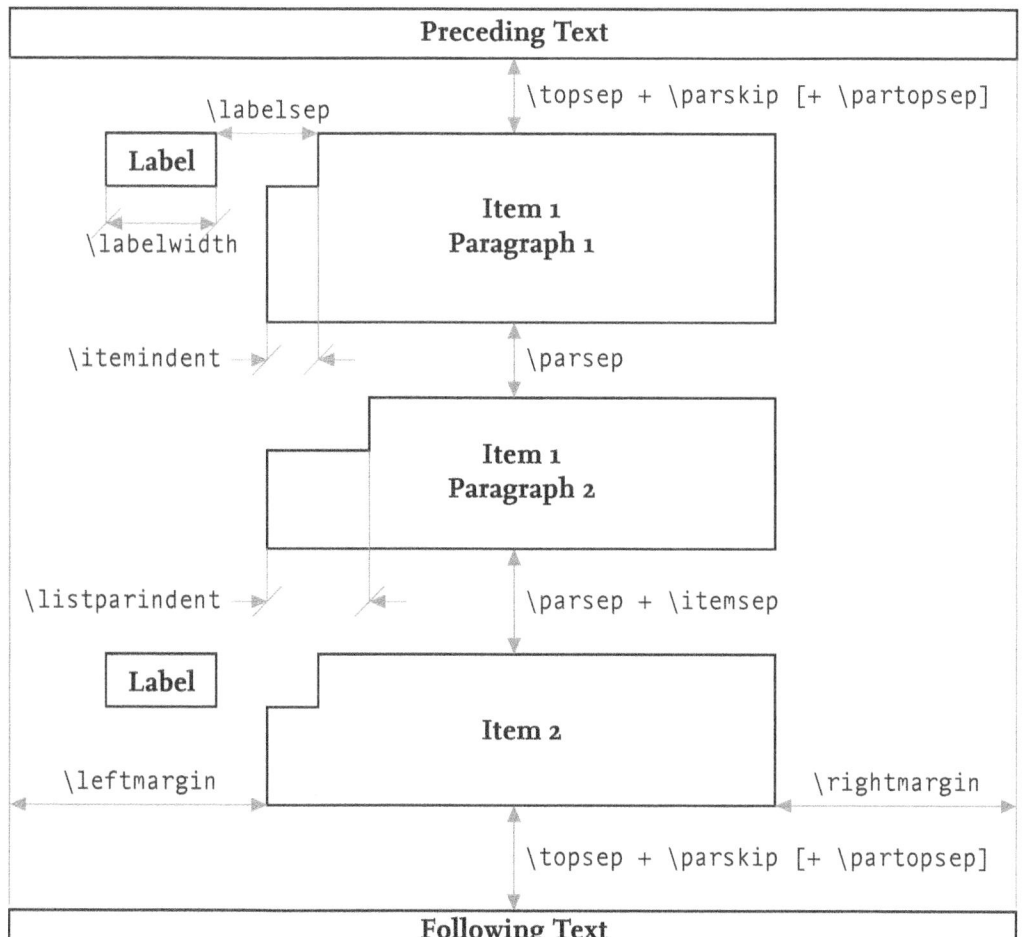

Figure 3.6
Lengths that affect list formatting

stretch depending on the lack or excess of vertical space on the page. By redefining the length commands inside the list environment you can change the appearance of the list.   ☑

Figure 3.7 presents an example of a user-defined list. The command \newcounter{ListCounter} defines a new counter called ListCounter. The spell List-\alph{ListCounter} typesets the label of each item as List- followed by the current value of the counter as a lowercase letter. Inside the second argument of the list environment, the command \usecounter{ListCounter} "uses" the counter, which basically adjusts the value of the counter for the next \item.

As part of LATEX's default mechanism all changes to counters and lengths inside an environment are local. This ensures that the counter ListCounter is reset to its original value upon leaving the environment.

Using the list environment over and over with the same arguments is not particularly useful and prone to errors. The \newenvironment command lets you define a new environment with an easier hassle-free interface. Figure 3.8 shows how you may implement the functionality of the list in Figure 3.7 as a user-defined environment.

Figure 3.7
A user-defined list

```
\newcounter{ListCounter}

\begin{list}
     {List-\alph{ListCounter}}
     {\usecounter{ListCounter}
      \setlength{\rightmargin}{0cm}
      \setlength{\leftmargin}{2cm}}
\item Hello.
\item World.
\end{list}
```

Figure 3.8
A user-defined environment for lists. In this example, the \newenvironment command takes three arguments. The first is the name of the new environment, the second argument determines what to do upon entering the environment, and the third what to do upon leaving the environment. The second argument opens a simple list environment and the third closes the list environment.

```
\newcounter{ListCounter}
...
% Define new environment:
\newenvironment
    {alphList}
    {\begin{list}
        {List-\alph{ListCounter}}
        {\usecounter{ListCounter}
         \setlength{\rightmargin}{0cm}
         \setlength{\leftmargin}{2cm}}}
    {\end{list}}
...
% Use new environment:
\begin{alphList}
   \item Hello.
   \item World.
\end{alphList}
```

The advantage of the user-defined environment is that it is easier to use and can be reused.

The first argument of \newenvironment is the name of the new environment. The second argument determines the commands that are carried out at the start of the environment. These are the commands that start the list environment. The last argument determines the commands that are carried out at the end on the environment. These commands end the list environment. More information about \newenvironment may be found in Chapter 11.

PART III

Tables, Diagrams, and Data Plots

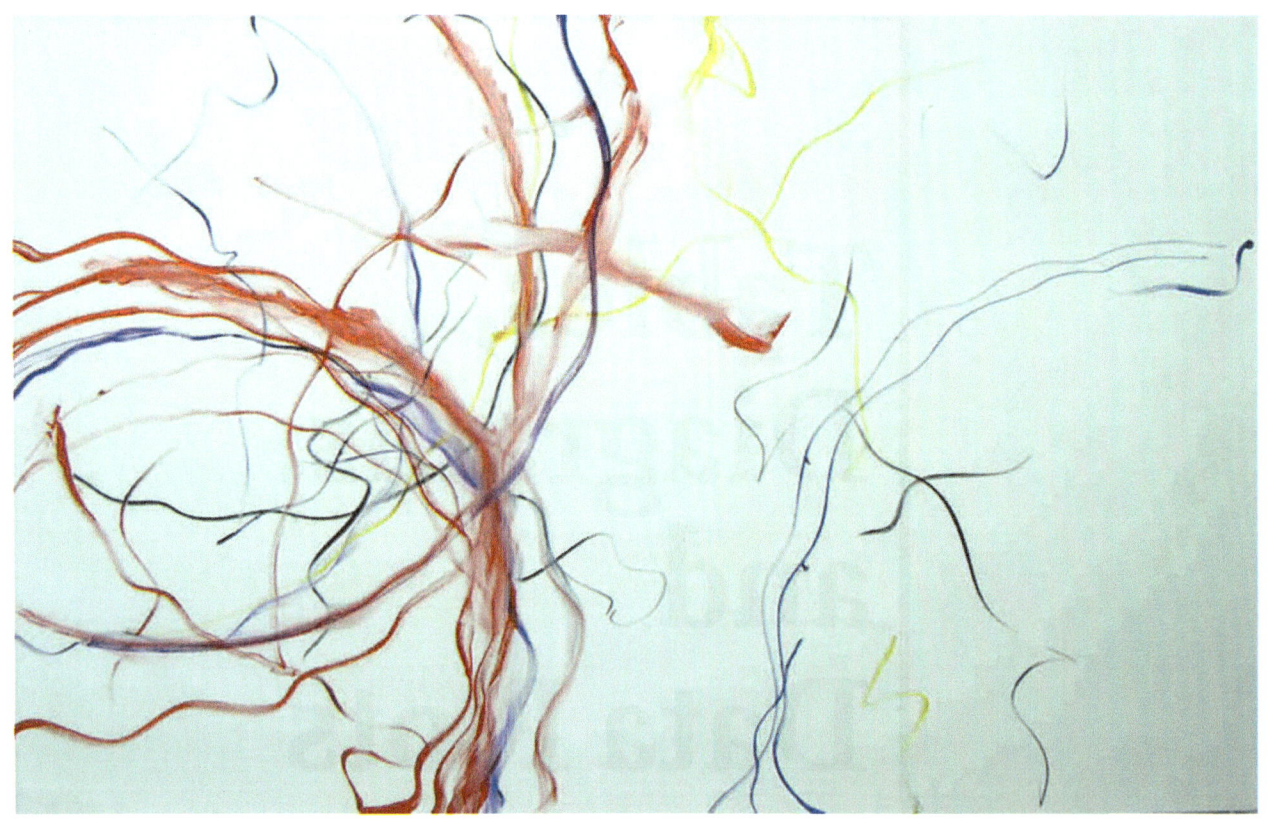

Oil and charcoal on canvas (24/01/08), 132 × 213 cm
Work included courtesy of Billy Foley
© Billy Foley (www.billyfoley.com)

CHAPTER 4
Presenting External Pictures

THIS CHAPTER is an introduction to presenting pictures that are stored in external files. Historically, this was an important mechanism for importing pictures. Since pictures are usually included as numbered figures, this chapter also provides an introduction to the `figure` environment and, more generally, *floating* environments.

The remainder of this chapter, is mainly based on [Carlisle 2005; Carlisle, and Rahtz 1999; Reckdahl 2006; Lamport 1994]. It starts by introducing the `figure` environment and continues by explaining how to include external pictures. This chapter is included mainly for completeness because Chapter 5 is an introduction to specifying pictures and diagrams with the `tikz` package. Furthermore, Chapter 7 shows how to present data plots using the `pgfplots` package. Readers not using external graphics are advised to only read the first two sections and skip the remainder of this chapter.

4.1 The `figure` Environment

The `figure` environment is usually used to present pictures, diagrams, and graphs. The environment creates a *floating* environment. *Floating* environments don't allow pagebreaks and they may "float" to a convenient location in the output document [Lamport 1994]. This mechanism gives LaTeX more freedom to choose better page breaks for the remaining text. For example, if there's not enough room left for a figure at the "current" position then LaTeX may fill up the remainder of the page with more paragraphs and put the figure on the next page. In this example the figure doesn't end up exactly where you intended it but the result is an aesthetically more pleasing document. However, it should be noted that you can also force the typesetting of a floating environment at the "current" position in the output file.

The body of a `figure` environment is typeset in a numbered figure. The `\caption` command may be used to define a caption of the figure.

LaTeX gives some control over the placement of floating figures, of floating tables, and other floats. For figures the placement is controlled with an optional argument of the `figure` environment. The same mechanism is used for the `table` environment, which is explained in Chapter 6. The optional argument, which controls the placement, may contain any combination of the letters t, b, p, h, and H, which are used as follows [Lamport 1994, page 197]:

t Put the float at the top of a page.

b Put the float at the bottom of a page.

p Put the float on a separate page with no text, but only figures, tables, and other floats.

h Put the float approximately at the current position (here). (This option is not available for double-column figures and figures in two-column format.)

H Put the float at the current position (Here). (This option is not available for double-column figures and figures in two-column format.)

The default value for the optional argument is tbp. LaTeX parses the letters in the optional argument from left to right and puts the figure at the position corresponding to the first letter for which it thinks the position is "reasonable." Good positions are the top of the page, the bottom of the page, or a page with floats only, because these positions do not disrupt the running text too much.

Inside the figure environment the command \caption defines a caption. The caption takes a *moving argument,* so fragile commands must be *protected.* Moving arguments and \protect are explained in Section 11.3.3. The regular argument defines the caption as it is printed in the figure and in the list of figures. If the regular argument gets too long then you may not want this text in the list of figures. In that case, you may add an optional argument, which is used to define a short alternative title for the list of figures. Within the regular argument of the \caption command, you may define a label for the figure with the \label command. This works as usual.

The following shows how to create a figure. Inside the figure environment you can put LaTeX statements to produce the actual figure. In this example the text 'Comparison of algorithms.' appears in the list of tables and the text 'Comparison. The dashed line ...' is used for the caption.

```
\begin{figure}[tbp]                              LaTeX Usage
    ⟨Insert LaTeX here to produce the figure.⟩
    \caption[Comparison of algorithms.]
            {Comparison. The dashed line ...
             \label{fig:comparison}}
\end{figure}
```

4.2 Special Packages

This section presents some packages that overcome some of LaTeX's limitations for floating environments.

4.2.1 *Floats*

LaTeX always forces the caption of floating environments on the same page as that of the environment The dpfloat package lets you create floating environments on consecutive pages. This may be useful, for

```
\begin{figure}[ptbh]
% Left-side part of float(possibly with \caption).
\begin{leftfullpage}
    ⟨Left part of float⟩
\end{leftfullpage}
\end{figure}
% Right-side part of float(possibly with \caption).
\begin{figure}[ptbh]
\begin{fullpage}
    ⟨Right part of float⟩
\end{fullpage}
\end{figure}
```

Figure 4.1

Using the dpfloat package

example, if your float is too large and you want the caption on the opposite page. Figure 4.1 presents an example.

4.2.2 *Legends*

Some documents distinguish between captions and *legends*. For such documents the caption of an environment consists of the name, the number, and a short title of the environment. For example, 'Figure 4.1: Using the dpfloat package.' The legend is a longer explanation of what's in the environment. LaTeX does not distinguish between captions and legends. The ccaption package overcomes this problem and provides a \legend command for legends. The package also provides support for caption placement and "sub-floats."

4.3 External Picture Files

Embedding pictures from external files is a common technique to creating graphics in documents. The best picture formats are vector graphics formats. The advantage of vector graphics is that they scale properly and always give the graphics a smooth appearance. Vector graphics formats that work well with LaTeX are eps and pdf.

Programs such as gnuplot may be used to generate graphs in vector graphics format from your data. A common practice is to generate complicated graphs with gnuplot and include them with LaTeX. This mechanism is relatively easy. However, gnuplot may not always have the right graph output style. Another problem with externally generated pictures is that they may not always give a consistent look and feel as a result of differences in fonts and scaling. The pgfplots package overcomes these problems. (This package is explained in Chapter 7.)

4.4 The graphicx Package

The graphicx package provides a command called \includegraphics that supports the inclusion of external graphics in an easy way.

Figure 4.2

Including an external graphics file with the includegraphics command. The input on the right results in the output on the left. The Doctor Fun picture is included with the kind permission of David Farley.

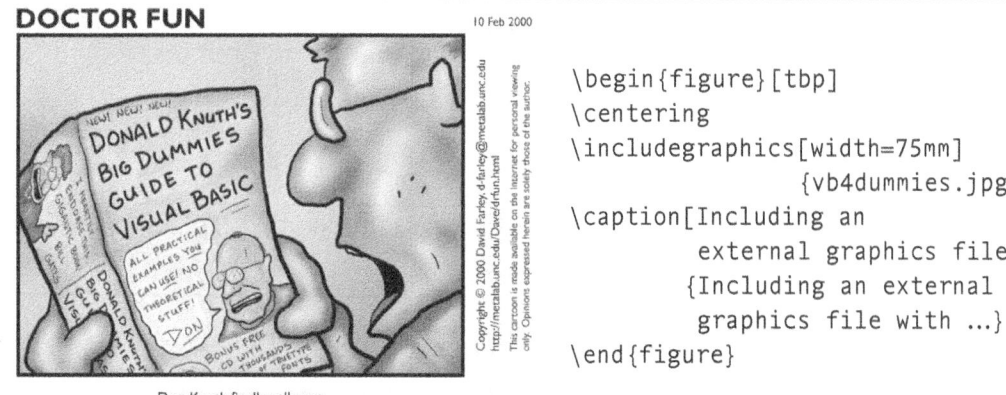

```
\begin{figure}[tbp]
\centering
\includegraphics[width=75mm]
                {vb4dummies.jpg}
\caption[Including an
         external graphics file]
         {Including an external
          graphics file with ...}
\end{figure}
```

\includegraphics[⟨key-value list⟩]{⟨file⟩}

This includes the external graphics file ⟨file⟩. The optional argument is a ⟨key⟩=⟨value⟩ list controlling the scale, size, rotation, and other aspects of the picture. The following describes some of the possible keys. Information about other ⟨key⟩=⟨value⟩ combinations may be found in the graphicx package documentation [Carlisle, and Rahtz 1999].

angle	The the rotation angle in degrees.
width	The width of the resulting picture. The width should be specified in a proper dimension, e.g., 5cm, 65mm, 3in, and so on. The height of the picture is scaled to match the given width.
height	The height of the resulting picture. This is the dual of the width key.
type	Specifies the file type. The file type is normally determined from the filename extension. ☑

Figure 4.2 shows an example of the \includegraphcs command. In this example, the command is used in the body of a figure. The picture is reproduced from the Dr Fun pages (http://www.ibiblio.org).

4.5 Setting Default Key Values

The graphicx package uses the keyval package to handle its ⟨key⟩= ⟨value⟩ pairs. The keyval package lets you define a default value for each key. The following is a short explanation. A full explanation may be found in the package documentation [Carlisle 1999b]. Basically, the command \setkeys{Gin}{⟨list⟩} sets the defaults. Here ⟨list⟩ is a comma-separated ⟨key⟩=⟨value⟩ list. The following example sets the default width to 6 cm.

```
\setkeys{Gin}{width=6cm}                          LaTeX Usage
```

After this command the width will be 6 cm by default. However, it is still possible to override this default width by providing an explicit width, an explicit height, or an explicit scale.

Sometimes it is nicer to specify a width and/or height as a fraction of the current paper dimensions. The following sets the default width to 90% of the text width.

```
\setkeys{Gin}{width=0.9\textwidth,}
```
LaTeX Usage

4.6 Setting a Search Path

By default \includegraphics searches the current directory for files. However, it is also possible to define a search path. The search path mechanism works similar to a Unix search path. The command \graphicspath{⟨directory list⟩} sets the search path to ⟨directory list⟩, which consists of a list of directories, each of which should be inside a brace pair. The following is an example which sets the search path to ./pdf/, ./eps. Notice the absence of commas in the list.

```
\graphicspath{{./pdf/}{./eps/}}
```
LaTeX Usage

4.7 Graphics Extensions

The kind of graphics extensions allowed by \includegraphics depends on the extension of your output file. The last argument of \includegraphics determines the name of the external graphics file. It is allowed to omit the file extension. When \includegraphics sees a filename without extension it will try to add a proper extension. The command \DeclareGraphicsExtensions{⟨extension list⟩} specifies the allowed file extensions. The argument ⟨extension list⟩ is a comma-separated list of extensions. The command works as expected. If an extension is omitted in the required argument of the \includegraphics command, ⟨extension list⟩ is searched from left to right. The process halts when an extension is found that "completes" the partial filename. The partial filename and the extension are used as the external graphics filename. Filenames without extensions are not allowed if you apply the command \DeclareGraphicsExtensions{}.

CHAPTER 5
Presenting Diagrams

THIS CHAPTER is an introduction to drawing diagrams/pictures using the tikz package, which is built on top of pgf. Here pgf is a platform- and format-independent macro package for creating graphics. The pgf package is smoothly integrated with TEX and LATEX. As a result tikz also lets you incorporate text and mathematics in your diagrams. The tikz package also supports the beamer class, which is used for creating incremental computer presentations. The beamer class is explained in Chapter 14.

The main purpose of this chapter is to whet the appetite. The presentation is mainly based on CVS version 2.00 (CVS2010-01-03) of pgf and tikz. The interested reader is referred to the excellent package documentation [Tantau 2010] for more detailed information.

This chapter starts with discussing the advantages and disadvantages of specifying diagrams. This is followed by a quick introduction to the tikzpicture environment and some drawing commands. Next there is a crash course on some of the more common and useful tikz commands. For most readers the material up to Section 5.12 should be enough to get by on a day-to-day basis. The remaining sections coordinate computations and advanced tikz commands. By the end of this chapter you should know how to draw maintainable, high-quality graphics consisting of basic shapes such as points, lines, circles, and trees.

5.1 Why Specify your Diagrams?

There are several advantages to tikz's approach of specifying your diagrams. It may take a while before you get the hang of tikz, but as you go along, you will find that learning the package gives you more control. If certain graphical entities serve a specific purpose then you can make them stand out by drawing them with a certain style. For example, you may decide to draw construction lines with thin, dashed, red lines. Drawing such lines like this gives your diagrams a consistent appearance. Furthermore, it makes it easy to recognise the parts of your diagrams. Of course, you can reuse existing styles and this simplifies the drawing. What is more, you can use stepwise refinement to develop your styles. As an overall result your diagrams will become more maintainable.

5.2 The `tikzpicture` Environment

The `tikz` package—`tikz` is an acronym of 'tikz ist kein Zeichen-programm'—provides commands and environments for specifying and "drawing" graphical objects in your document. The package is smoothly integrated with TeX and LaTeX, so graphical objects also can be text. What is more, the things you specify/draw may have attributes. For example, tree nodes have coordinates and may have parts such as children, grandchildren, and so on. The package also supports mathematical and object orientated computations.

Drawing with `tikz` may be done in different ways. To simplify matters we shall do most of our drawing inside a `tikzpicture` environment.

Each `tikzpicture` environment results in a picture containing what is in the bounding box of the environment. Only the relative positions of the coordinates inside a `tikzpicture` matter. For example, a `tikzpicture` consisting of a 2×2 square that is drawn at coordinate $(1,2)$ in the `tikzpicture` results in the same graphic on your page as a `tikzpicture` consisting of a 2×2 square that is drawn at coordinate $(0,0)$ in your `tikzpicture`. All *implicit* units inside a `tikzpicture` are in centimetres. Scaling a `tikzpicture` is done by passing an option of the form `scale=number` to the `tikzpicture` environment. This kind of scaling only applies to the actual coordinates but *not* to line thicknesses, font sizes, and so on. This makes sense, because you would not want, say, the font in your diagrams to be of a different kind than the font in your running text. The package also supports other top-level options.

The following draws a 0.4×0.2 crossed rectangle: ⊠.

```
                                                          LaTeX Usage
The following draws
 a $0.4 \times 0.2$ crossed rectangle:
\begin{tikzpicture}
\draw (0.0,0.0) rectangle (0.4,0.2);
\draw (0.0,0.0) -- (0.4,0.2);
\draw (0.0,0.3) -- (0.4,0.0);
\end{tikzpicture}\,.
```

Of course this example violates every rule in the maintainability book. For example, what if the rectangle's size were to change, what if its position were to change, what if its colour were to change, ...?

Fortunately, `tikz` provides a range of commands and techniques for maintaining your diagrams. One of the cornerstones is the ability to label nodes and coordinates and use the labels to construct other nodes and shapes. In addition the package supports hierarchies. Parent settings may be inherited by descendants in the hierarchy.

5.3 The `\tikz` Command

The `\tikz` command is useful for small in-line diagrams. The following explains how it works.

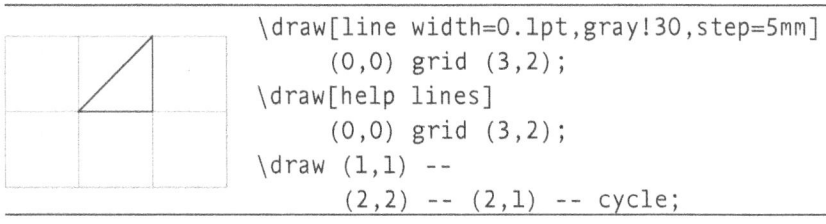

```
\draw[line width=0.1pt,gray!30,step=5mm]
     (0,0) grid (3,2);
\draw[help lines]
     (0,0) grid (3,2);
\draw (1,1) --
     (2,2) -- (2,1) -- cycle;
```

Figure 5.1

Drawing a grid

\tikz[⟨options⟩]{⟨commands⟩}

This works just as \begin{tikzpicture}[⟨options⟩]⟨commands⟩\end {tikzpicture}. ☑

\tikz[⟨options⟩] ⟨command⟩;

If ⟨command⟩ is a single command then this is equivalent to \tikz [⟨options⟩]{⟨command⟩;}. ☑

From now on all examples are implicitly inside a tikzpicture environment unless options are passed to the environment.

5.4 Grids

A grid relates the positions of what's in the picture. Grids are also useful when you are developing a picture. The following shows two ways to draw a grid. The former way is easier, but it is expressed in terms of the second, more general, notation.

\draw[⟨options⟩] ⟨start coordinate⟩ grid ⟨end coordinate⟩;

This draws a grid from ⟨start coordinate⟩ to ⟨end coordinate⟩. The optional argument may be used to control the style of the grid.

The option step=⟨dimension⟩ is used for setting the distance between the lines in the grid. There are also directional versions xstep = ⟨dimension⟩ and ystep = ⟨dimension⟩ for setting the distances in the x- and y-directions. ☑

\path[⟨path options⟩] ... grid[⟨options⟩] ⟨coordinate⟩ ... ;

This adds a grid to the current path from the current position in the path to ⟨coordinate⟩. To *draw* the grid, the option draw is required as part of ⟨path options⟩. ☑

Figure 5.1 demonstrates how to draw a basic 3 × 2 grid, relative to the origin. The grid consists of two superimposed grids, the coarser of which is drawn on top of the other. The option gray!30 in the style of the fine grid defines the colour for the grid: you get it by mixing 30% grey and 70% white.

The option help lines draws lines in a subdued colour. This book redefined the style to make the lines thin and to set the color to a combination of 50% black and 50% white. This was done with the command \tikzset{help lines/.style={thin,color=black!50}}. Styles are explained in Section 5.17.

5.5 Paths

Inside a tikzpicture environment everything is drawn by starting a *path* and by *extending* the path. Paths are constructed using the \path

Figure 5.2

Creating a path

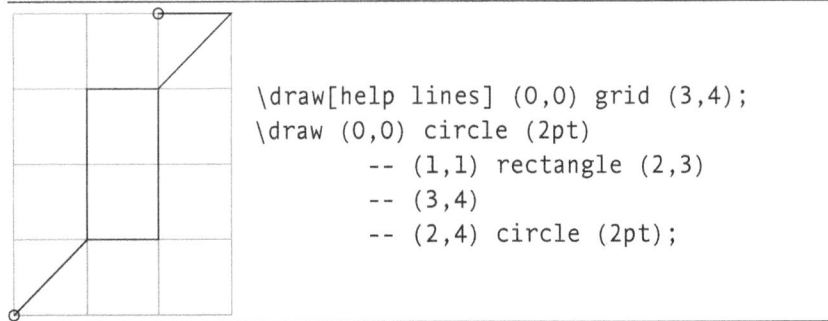

```
\draw[help lines] (0,0) grid (3,4);
\draw (0,0) circle (2pt)
        -- (1,1) rectangle (2,3)
        -- (3,4)
        -- (2,4) circle (2pt);
```

command. In its basic form, a path is started with a coordinate that becomes the *current* coordinate of the path. Next the path is extended with other coordinates, line segments, *nodes* or other shapes. Line segments may be straight line segments or *cubic spline segments,* which are also known as *cubic splines.* Cubic splines are explained in Section 5.7. Each line segment extension operation adds a line segment starting at the current coordinate and ending at another coordinate. Path extension operations may update the current coordinate.

The optional argument of the \path command is used to control if, and how the path should be drawn. Adding the option draw forces the drawing of the path. By default the path is *not* drawn. A semicolon indicates the end of the path:

```
\path[draw] (1,0) -- (2,0);
\path       (0,0) -- (3,0);
```
LaTeX Usage

The first \path command in this tikzpicture draws a line segment from $(1, 0)$ to $(2, 0)$. The second \path command draws an invisible line segment. Both line segments are considered part of the picture, so the picture has a width of 3 cm.

The command \draw is a shorthand for \path[draw]. The tikz package has many shorthand notations like this.

Figure 5.2 draws a path that starts at position $(0, 0)$. First the path is extended by adding a circle. Next the path is extended with a line segment leading to $(1, 1)$. Next it is extended with a rectangle, and so on. Except for the circle extension operation, each operation changes the current position of the path.

5.6 Coordinate Labels

Maintaining complex diagrams defined entirely in terms of absolute coordinates is virtually impossible. Fortunately, tikz provides many techniques that help maintain your diagrams. One of these techniques is that tikz lets you define coordinate labels and use the resulting labels instead of the coordinates.

You define a coordinate label by writing coordinate(⟨label⟩) after the coordinate. Defining coordinates this way is possible at (almost) any point in a path. Once the label of a coordinate is defined, you can use (⟨label⟩) as a coordinate. The following, which draws a crossed rectangle (⊠), demonstrates the mechanism. It is not intended to excel in terms of maintainability.

```
The following, which draws a crossed rectangle          LaTeX Usage
(\begin{tikzpicture}
  \draw (0.0,0.0) coordinate(lower left)
        -- (0.4,0.2) coordinate(upper right);
  \draw (0.0,0.2) -- (0.4,0.0);
  \draw (lower left) rectangle (upper right);
\end{tikzpicture}), demonstrates the mechanism.
```

5.7 Extending Paths

As explained before, paths are constructed by *extending* them. There are several different kinds of path extension operations. The majority of these extension operations modify the current coordinate, but some don't. In the remainder of this section it is therefore assumed that an extension operation modifies the current coordinate *unless* this is indicated otherwise. For the moment it is assumed that none of the coordinates are relative or incremental coordinates, which are explained in Section 5.15.1.

The remainder of this section presents some examples. To save space, operations are explained in terms of the \path operation. However, \path does not draw anything without the option draw. Rather than giving examples in terms of \path[draw], they are given for \draw, which is equivalent. The following are the common extension operations.

\path ... ⟨coordinate⟩ ...;

This is the move-to operation, which adds the coordinate ⟨coordinate⟩ to the path. This operation makes ⟨coordinate⟩ the current coordinate of the path. The following example uses three move-to operations. The first move-to operation defines the lower left corner of the grid. The remaining move-to operations define the starts of two line segments.

```
\draw[help lines] (0,0) grid (3,2);
\draw (0,0) -- (1,1)
      (2,1) -- (3,2);                              ☑
```

\path ... -- ⟨coordinate⟩ ...;

This is the line-to operation, which adds a straight line segment to the path. The line segment is from the current coordinate and ends in ⟨coordinate⟩.

```
\draw[help lines] (0,0) grid (3,2);
\draw (0,0) -- (1,1) --
      (2,0) -- (3,2);                              ☑
```

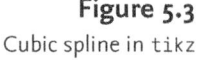

Figure 5.3
Cubic spline in tikz

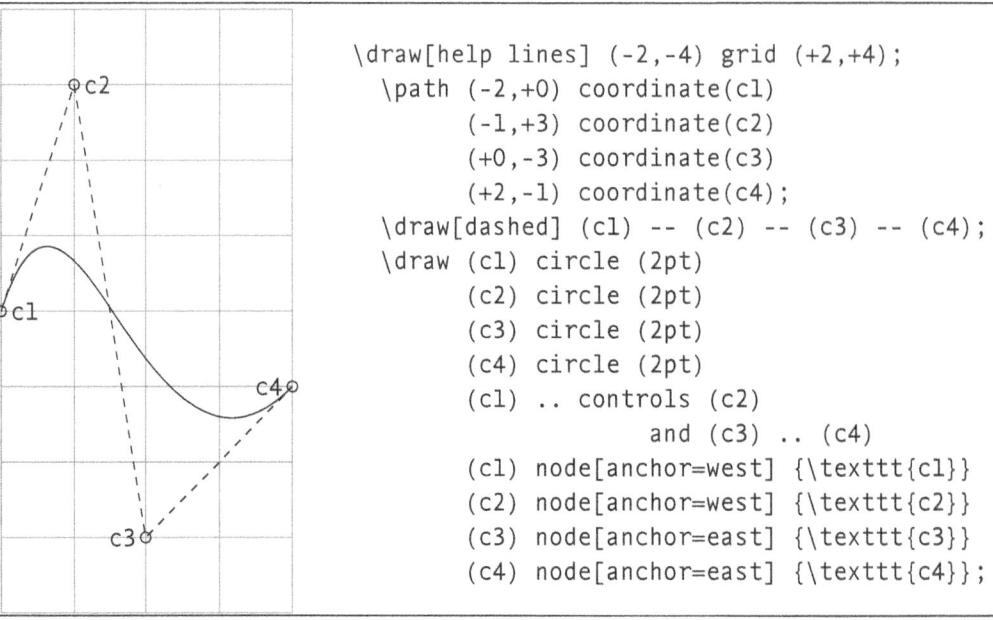

```
\draw[help lines] (-2,-4) grid (+2,+4);
  \path (-2,+0) coordinate(c1)
        (-1,+3) coordinate(c2)
        (+0,-3) coordinate(c3)
        (+2,-1) coordinate(c4);
\draw[dashed] (c1) -- (c2) -- (c3) -- (c4);
\draw (c1) circle (2pt)
      (c2) circle (2pt)
      (c3) circle (2pt)
      (c4) circle (2pt)
      (c1) .. controls (c2)
                    and (c3) .. (c4)
      (c1) node[anchor=west] {\texttt{c1}}
      (c2) node[anchor=west] {\texttt{c2}}
      (c3) node[anchor=east] {\texttt{c3}}
      (c4) node[anchor=east] {\texttt{c4}};
```

\path ... .. controls ⟨coordinate₁⟩ and ⟨coordinate₂⟩ .. ⟨coordinate₃⟩ ...;

This is the curve-to operation, which adds a *cubic Bézier spline segment* to the path. The start point of the curve is the current point of the path. The end point is ⟨coordinate₃⟩, and the control points are ⟨coordinate₁⟩ and ⟨coordinate₂⟩.

Figure 5.3 demonstrates the operation. The curve starts at c1 and ends at c4. The control points are given by c2 and c3. The tangent of the spline segment at c1 is equal to the tangent of the line segment c1 -- c2. Likewise, the tangent at c4 is given by the tangent of the line segment c3 -- c4.

This makes cubic Bézier splines a perfect candidate for approximating complex curves as a sequence of spline segments. By properly choosing the start point, the end point, and the control points of the segments, you can enforce continuity both in the curves *and* the first derivative. For example, if the end point of segment s_1 coincides with the start point of the segment s_2 and the end point of s_1 is on the line through the last control point of s_1 and the first control point of s_2, then this guarantees continuity of the first derivative. (As a matter of fact, it is also possible to ensure continuity in the second derivative.) Notice that the start, end, and control points need not be equidistant, nor need the start and end point lie on a horizontal line. ☑

\path ... .. controls ⟨coordinate₁⟩ .. ⟨coordinate₂⟩ ...;

This is also a curve-to operation. It is equivalent to the operation ... .. controls ⟨coordinate₁⟩ and ⟨coordinate₁⟩ .. ⟨coordinate₂⟩ .... ☑

\path ... -- cycle ...;

This is the cycle operation, which closes the current path by adding a straight line segment from the current point to the most recent destination point of a move-to operation. The cycle operation has three applications. First it *closes* the path, which is required if you wish to *fill* the path with a colour. Second, it connects the start and end line segments in the path. Third, it avoids the need to reference the start point of the path.

```
\draw (0,0) -- (1,1)
      (2,0) -- (3,0) --
      (3,1) -- cycle;
```

\path ... -| ⟨coordinate⟩ ...;

This operation is equivalent to two line-to operations connecting the current coordinate and ⟨coordinate⟩. The first operation adds a horizontal and the second a vertical line segment.

```
\tikz \draw (0.0,0.0) -| (2.0,0.5)
            (1.0,1.0) -| (3.0,0.0);
```

\path ... |- ⟨coordinate⟩ ...;

This operation is also equivalent to two line-to operations connecting the current coordinate and ⟨coordinate⟩. This time, however, the first operation adds a vertical and the second a horizontal line segment.

```
\tikz \draw (0.0,0.0) |- (2.0,1.0)
            (1.0,0.5) |- (3.0,0.0);
```

\path ... rectangle ⟨coordinate⟩ ...;

This is the rectangle operation, which adds a rectangle to the path. The rectangle is constructed by making the current coordinate and ⟨coordinate⟩ the lower left and upper right corners of the rectangle. Which coordinate determines which corner depends on the values of the coordinates.

```
\begin{tikzpicture}
\draw (0,0) rectangle (1,1)
            rectangle (3,2);
\end{tikzpicture}
```

\path ... circle (⟨radius⟩) ...;

This is the circle operation, which adds a circle to the path. The centre of the circle is given by the current coordinate of the path and its radius is the dimension ⟨radius⟩. This operation does *not* change the current coordinate of the path.

```
\tikz \draw (0,0) circle (2pt)
            rectangle (3,1)
            circle (4pt);
```

\path ... ellipse(⟨half width⟩ and ⟨half height⟩) ...;

This is the ellipse operation, which adds an ellipse to the path. The centre of the ellipse is given by the current coordinate. This operation does *not* change the current coordinate of the path.

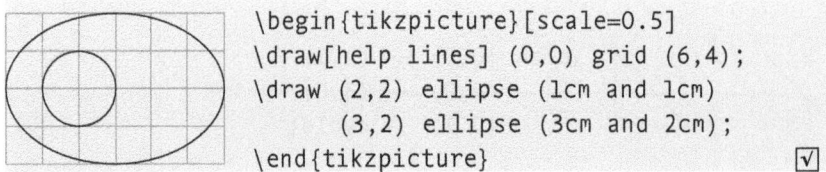

```
\begin{tikzpicture}[scale=0.5]
\draw[help lines] (0,0) grid (6,4);
\draw (2,2) ellipse (1cm and 1cm)
      (3,2) ellipse (3cm and 2cm);
\end{tikzpicture}
```
☑

\path … arc ($\alpha:\beta:r$) …;

This is the arc operation, which adds an arc to the path. The arc starts at the current point, p. The arc is determined by a circle with radius r. The centre of the circle, c, is determined by the equation $p = c + r \times (\cos\alpha, \sin\alpha)$. The end point of the arc is given by $c + r \times (\cos\beta, \sin\beta)$. The arc is drawn in counterclockwise direction from the start point to the end point, which becomes the new current coordinate of the path. The following illustrates the construction. Only the upper half of the circle is drawn. The resulting arc is drawn with a continuous line.

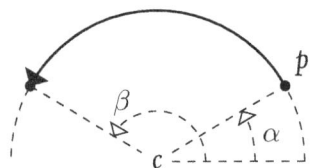

In the following example, the draw option -> results in a line that is drawn as an arrow.

```
\begin{tikzpicture}
\draw[help lines] (0,0) grid (3,2);
\draw[dashed] (1,1) circle (1cm);
\draw (1,2) coordinate(a) circle (2pt)
      (2,1) coordinate(b) circle (3pt)
      (1,0) coordinate(c) circle (4pt);
\draw[->,thick] (a) arc (90:180:1cm);
\draw[->,thick] (b) arc (0:45:1cm);
\draw[->,thick] (c) arc (270:225:1cm);
\end{tikzpicture}
```
☑

\path … arc ($\alpha:\beta:w$ and h) …;

This is the elliptical arc operation, which adds an ellipse segment to the path. The construction of the ellipse segment is similar to the construction of the arc segment. The value h is half the hight of the ellipse and the value w is half the width of the ellipse. ☑

5.8 Actions on Paths

So far most of our examples have used the default path style. This may not always be what you want. For example, you may want to draw a line in a certain colour, change the default line width, fill a shape with a colour, and so on. In tikz terminology you achieve this with *path actions*, which are operations acting on an existing path. You first construct the path and then apply the action. At the basic level the command \draw is defined in terms of an action on a path: the action results in the path being drawn. As pointed out before \draw is a shorthand for \path[draw].

The following are some other shorthand commands that are defined in terms of path actions inside the tikzpicture environment.

\draw

This is a shorthand for \path[draw].

```
\draw (0,0) -- (3,0);                    ☑
```

\fill

This is a shorthand for \path[fill].

```
\fill[gray] (0,0) rectangle (3,0.5);     ☑
```

\filldraw

This is a shorthand for \path[filldraw].

```
\filldraw[fill=gray,draw=black]
         (0,0) rectangle (3,0.5);        ☑
```

\shade

This is a shorthand for \path[shade]. Shading paths is possible in many ways. The reader is invited to read the pgf manual [Tantau 2010] for further information. The following is an example.

```
\shade[left color=black,right color=gray]
      (0,0) rectangle (3,0.5);           ☑
```

\shadedraw

This is a shorthand for \path[shadedraw].

```
\shadedraw[left color=black,
           right color=white,
           draw=gray]
          (0,0) rectangle (3,0.5);       ☑
```

5.8.1 *Colour*

The tikz package knows several colours. Some colours are inherited from the xcolor package [Kern 2007]. Table 5.1 depicts some of them.

There are several techniques to define a new name for a colour.

\definecolor{⟨name⟩}{rgb}{⟨red⟩,⟨green⟩,⟨blue⟩}

This defines a new colour called ⟨name⟩ in the rgb model. The colour

Table 5.1

The xcolor colours

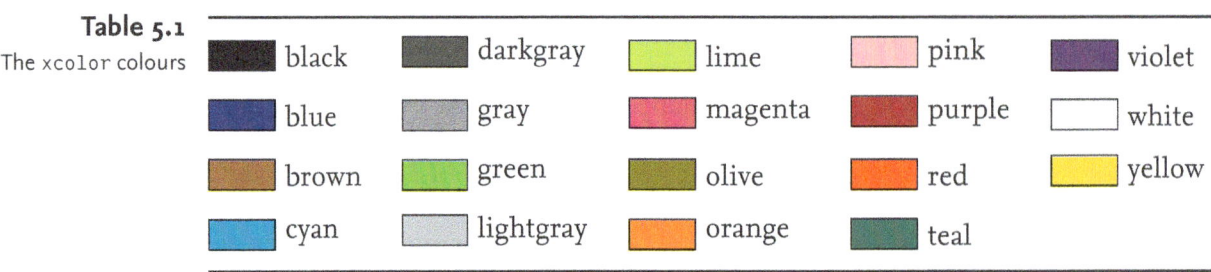

■	black	■	darkgray	■	lime	■	pink	■	violet
■	blue	■	gray	■	magenta	■	purple	□	white
■	brown	■	green	■	olive	■	red	■	yellow
■	cyan	■	lightgray	■	orange	■	teal		

is the result of combining ⟨red⟩ parts red, ⟨green⟩ parts green, and ⟨blue⟩ parts blue. All parts are reals in the interval [0 : 1].　☑

`\definecolor{⟨name⟩}{gray}{⟨ratio⟩}`

　　　This defines a colour called ⟨name⟩ that has a ⟨ratio⟩ grey part in the gray model. The value of ⟨ratio⟩ is a real in the interval [0 : 1].　☑

`\colorlet{⟨name⟩}{⟨colour⟩!⟨percentage⟩}`

　　　This defines a new colour called ⟨name⟩ that is the result of mixing ⟨percentage⟩% ⟨colour⟩ and (100 − ⟨percentage⟩)% white. Here ⟨colour⟩ should be the name of an existing colour.　☑

`\colorlet{⟨name⟩}{⟨colour₁⟩!⟨percentage⟩!⟨colour₂⟩}`

　　　This defines a new colour called ⟨name⟩ that is the result of mixing ⟨percentage⟩% ⟨colour₁⟩ and (100 − ⟨percentage⟩)% ⟨colour₂⟩, Here ⟨colour₁⟩ and ⟨colour₂⟩ should be existing colours.　☑

　　　Kern [2007] provides further information about defining colours with the xcolor package.

　　　Some path actions also let you define a colour. For example, you may draw a path with the given colour. There are different ways to control the colour. The option color determines the colour for drawing and filling, and the colour of text in nodes. (Nodes are explained in Section 5.9.)

　　　You may set the colour of the whole tikzpicture or set the colour of a given path action. Setting the colour of the whole picture is done by passing a color=⟨colour⟩ option to the environment. Setting the colour of a path action is done by passing the option to the \path command (or derived shorthand commands). The following is an example that draws three lines: one in red, one in green, and one in 50% cyan and 50% red.

```
\begin{tikzpicture}[color=red]
\draw                   (0,3) -- (2,3);
\draw[color=green]      (0,2) -- (2,2);
\draw[color=cyan!50!red] (0,1) -- (2,1);
\end{tikzpicture}
```

It is usually possible to omit the color= part when you specify colour options.

```
\begin{tikzpicture}[gray]
\draw[orange!80!teal] (0,0) -- (2,0);
\end{tikzpicture}
```

```
\draw[dash pattern=on 4mm off 1mm on 4mm off 2mm]
      (0,0.5) -- (2,0.5);
\draw[dash pattern=on 3mm off 2mm on 3mm off 3mm]
      (0,0.0) -- (2,0.0);
```

Figure 5.4

Using a dash pattern

```
\begin{tikzpicture}[dash pattern=on 3mm off 2mm]
\draw[dash phase=3mm] (0,0.5) -- (2,0.5);
\draw[dash phase=2mm] (0,0.0) -- (2,0.0);
\end{tikzpicture}
```

Figure 5.5

Using a dash phase

5.8.2 Drawing the Path

As already mentioned, the draw option draws the path in the default colour. You may also provide an explicit colour with draw=⟨colour⟩.

```
\draw[draw=gray] (0,1) -- (2,1);
```

5.8.3 Line Width

There are several path actions affecting the line style, i. e., the style that determines the line width, the line cap, and the line join. The following command lets you set the line width to an explicit value.

line width=⟨dimension⟩

This sets the line width to ⟨dimension⟩.

```
\draw[line width=8pt]
      (0,0) -- (2,4pt);
```
☑

5.8.4 Dash Patterns

The drawing of lines also depends on the *dash pattern* and *dash phase* settings. The dash pattern determines a basic pattern for the line that is repeated cyclicly. The dash phase shifts the dash pattern. By default the dash pattern is solid. The following shows the relevant path actions that affect dash patterns.

dash pattern=⟨pattern⟩

This sets the dash pattern to ⟨pattern⟩. The syntax for ⟨pattern⟩ is the same as in METAFONT. The ⟨pattern⟩ is cyclic pattern of lengths that determine when the line is drawn (is on) and when it is not drawn (is off). You usually write the lengths in multiples of points (pt). Figure 5.4 shows an example that uses millimetres. ☑

dash phase=⟨dimension⟩

This shifts the dash phase by ⟨dimension⟩. An example is presented in Figure 5.5. ☑

Table 5.2
Line width and dash pattern styles

	Line Styles		Dash Patterns	
Name	**Width**	**Example**	**Name**	**Example**
ultra thin	0.1 pt		loosely dotted	
very thin	0.2 pt		dotted	
thin	0.4 pt		densely dotted	
semithick	0.6 pt		solid	
thick	0.8 pt		loosely dashed	
very thick	1.2 pt		dashed	
ultra thick	1.6 pt		densely dashed	

5.8.5 *Predefined Styles*

Hard-coding a line width or a dash pattern command is not always a good idea. It is usually better to define a style for a certain line width, for a dash style, or a combination of the two. The advantages of doing this are that you only have to define the style once and can use it several times. Using styles gives you a consistent appearance for the resulting lines, and if you want to make a global change to the style then you only have to make one change in your LaTeX file. Section 5.17 explains how to define your own styles.

Table 5.2 lists the names of of some predefined line width and dash pattern styles. For line width styles, the table also lists the corresponding line width in points, and an example of the resulting line. The default line width is thin. The default dash pattern is solid.

5.8.6 *Line Cap and Join*

The drawing of a path depends on several parameters. The *line cap* determines how lines start and end. The *line join* determines how line segments are joined.

`line cap=⟨style⟩`

This sets the line cap style to ⟨style⟩. There are three possible values for ⟨style⟩: round, rect, and butt. The following shows the line cap options in action.

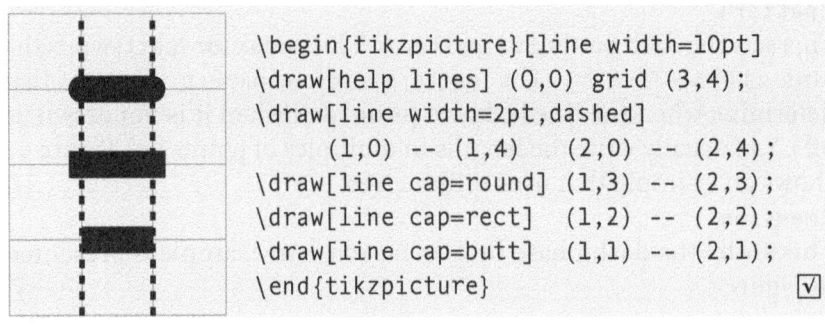

```
\begin{tikzpicture}[line width=10pt]
\draw[help lines] (0,0) grid (3,4);
\draw[line width=2pt,dashed]
      (1,0) -- (1,4)   (2,0) -- (2,4);
\draw[line cap=round] (1,3) -- (2,3);
\draw[line cap=rect]  (1,2) -- (2,2);
\draw[line cap=butt]  (1,1) -- (2,1);
\end{tikzpicture}
```

```
\begin{tikzpicture}
      [line width=8pt,line join=miter]
\draw (0,0) -- (0.25,2) -- (0.5,0);
\draw[miter limit=8]
      (1,0) -- (1.25,2) -- (1.5,0);
\end{tikzpicture}
```

Figure 5.6
Using the miter option

line join=⟨style⟩

This sets the line join style to ⟨style⟩. There are three possible values for ⟨style⟩: round, miter, and bevel.

```
\begin{tikzpicture}[line width=8pt]
\draw[line join=round]
      (0.0,.8)--(0.3,.0)--(0.6,.8);
\draw[line join=miter]
      (0.9,.0)--(1.2,.8)--(1.5,.0);
\draw[line join=bevel]
      (1.8,.8)--(2.1,.0)--(2.4,.8);
\end{tikzpicture}                       ☑
```

miter limit=⟨fraction⟩

This avoids sharp-angled miter joins that protrude too far beyond the joining point. It does this by posing a limit on how far the miter join may protrude the joining point. If the join protrudes beyond the limit then the join style is changed to bevel. The limit is equal to the product of ⟨fraction⟩ and the line width. Figure 5.6 shows an example.   ☑

5.8.7 Arrows

Arrows are also drawn using path actions. The following explains how to draw them.

arrows=⟨head₁⟩-⟨head₂⟩

This adds an arrow head to the start and to the end of the path. You may also omit the arrows= and use the shorthand notation ⟨head$_1$⟩-⟨head$_2$⟩. The arrow head at the start is determined by ⟨head$_1$⟩. The other arrow head is determined by ⟨head$_2$⟩. If you omit ⟨head$_1$⟩ then the arrow head at the start of the path isn't drawn. Likewise, omitting ⟨head$_2$⟩ omits the other arrow head. The following demonstrates the mechanism for the default arrow head types < and >. Table 5.3 list more arrow head styles. Some of the styles are provided by the tikz library arrows.

```
\draw[->]  (0,1.0) -- (2,1.0);
\draw[<-]  (0,0.5) -- (2,0.5);
\draw[<->] (0,0.0) -- (2,0.0);     ☑
```

Table 5.3

Some available arrow head types. The arrows in the upper part of the table are predefined. The arrows in the lower part of the table are provided by the tikz library arrows.

					Predefined
Style	**Arrow**	**Style**	**Arrow**	**Style**	**Arrow**
stealth	⟶	to	⟶	latex	⟶
space	—				

					Provided by arrows
Style	**Arrow**	**Style**	**Arrow**	**Style**	**Arrow**
open triangle 90	⟶▷	triangle 90	⟶▶	angle 90	⟶→
open triangle 60	⟶▷	triangle 60	⟶▶	angle 60	⟶→
open triangle 45	⟶▷	triangle 45	⟶▶	angle 45	⟶→
open diamond	⟶◇	diamond	⟶◆	o	⟶○
open square	⟶□	square	⟶■	*	⟶●

>=⟨end arrow type⟩

This redefines the *default* end arrow head style >. As already mentioned, some existing arrow head styles are listed in Table 5.3. Some of these arrow head types are provided by the tikz library arrows. Most styles in the table also have a "reversed style," for example latex reversed, which just changes the direction of the latex arrow head. The library may be loaded with the command \usetikzlibrary{arrows}, which should be in the preamble of your document. The following is a small example.

```
\draw[>=o,<->]      (0,1.0) -- (2,1.0);
\draw[>=*,<-]       (0,0.5) -- (2,0.5);
\draw[>=latex,->]   (0,0.0) -- (2,0.0);   ☑
```

5.8.8 Filling a Path

Not only can you draw paths but also can you fill them or draw them with one colour and fill them with a different colour. The only requirement is that the path be closed. Closing a path is done with the cycle annotation. The following are the relevant commands.

\path[fill=⟨colour⟩] ⟨paths⟩;

This fills each path in ⟨paths⟩ with the colour ⟨colour⟩. Unclosed paths are closed first. It is also allowed to use color=⟨colour⟩. Finally, the option fill on its own fills the paths with the last defined value for fill or for color.

```
\begin{tikzpicture}[scale=0.4,fill=gray]
\path[fill]
     (0,0) rectangle (1,1);
\path[fill=black!30]
     (2,0) -- (3,0) -- (3,1) -- cycle;
\path[fill,color=gray]
     (4,0) -- (5,0) -- (5,1);
\end{tikzpicture}                      ☑
```

`\fill[⟨options⟩] ⟨paths⟩;`

The command `\fill` on its own works just as `\path[fill=⟨colour⟩]`, where ⟨colour⟩ is the last defined value for `fill` or for `color`. Using `\fill` with options works as expected. The options are passed to `\path` and the paths in ⟨paths⟩ are filled.

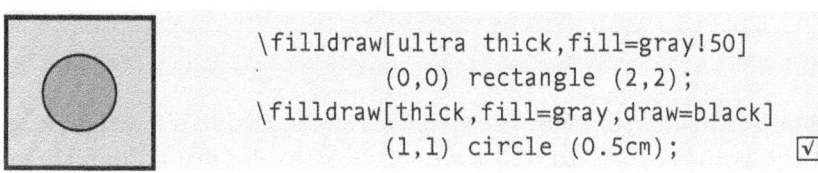

```
\begin{tikzpicture}[scale=0.4,fill=gray]
\fill[color=black!50]
      (0,0) -- (1,0) -- (1,1);
\fill[fill=black]
      (0,1) -- (0,2) -- (1,2) -- cycle;
\fill[gray!50]
      (2,0) -- (3,0) -- (3,1) -- cycle;
\fill(2,1) -- (2,2) -- (3,2);
\end{tikzpicture}                      ☑
```

`\filldraw[options] ⟨paths⟩;`

The command `\filldraw` fills and draws the path. The style `draw` determines the drawing colour and the style `fill` determines the filling colour. Both styles may be set in the optional argument.

```
\filldraw[ultra thick,fill=gray!50]
          (0,0) rectangle (2,2);
\filldraw[thick,fill=gray,draw=black]
          (1,1) circle (0.5cm);        ☑
```

5.8.9 *Path Filling Rules*

There are two options that control how overlapping paths are filled. Basically, these rules determine which points are inside a given path and this determines how the path is filled. By cleverly using the two options and by making paths overlap you can construct holes in the filled areas. The following explains the rules.

`nonzero` rule The `nonzero` rule is the default rule to determine which points are inside the path. To determine if a point, p, is inside a collection of paths, let c^+ be the number of clockwise draw paths the point is in, and let c^- be the number of anticlockwise draw paths the point is in. Then p is considered inside the collection of paths if $c^+ \neq c^-$. Stated differently, p is inside if $c^+ - c^- \neq 0$, hence the name `nonzero` rule.

To complicate matters, closed paths may "overlap" themselves and this may result in points that are in clockwise as well as anticlockwise sub-paths. To determine if a point, p, is inside the paths, let ℓ be a semi-infinite line originating at p. Then p is inside the paths if the number of times ℓ crosses a clockwise drawn line differs from the number of times ℓ crosses an anticlockwise line. Figure 5.7 shows an example. The example does *not* have self-overlapping paths.

`even odd` rule The `even odd` rule is the other rule. According to this rule a

Figure 5.7

Using the `nonzero` rule. The fill involves three rectangular sub-paths. The red sub-path is drawn clockwise; the other sub-paths are drawn anticlockwise. For the nonzero rule a point, p is filled if $c^+ \neq c^-$, where c^+ (c^-) is the number of clockwise (anticlockwise) shapes p is in.

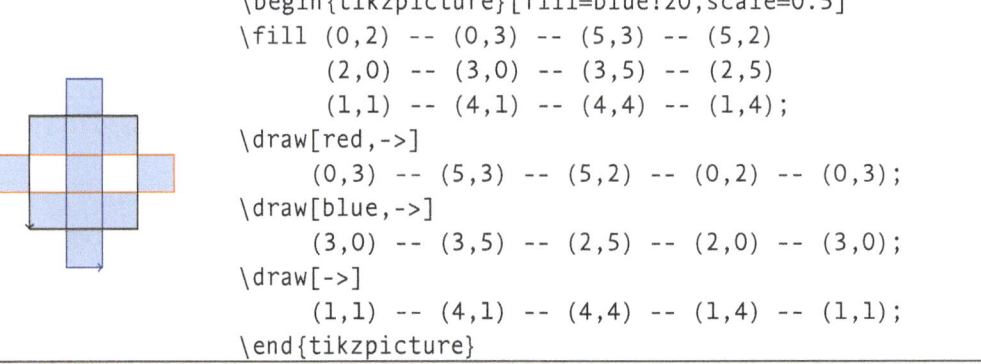

```
\begin{tikzpicture}[fill=blue!20,scale=0.5]
\fill (0,2) -- (0,3) -- (5,3) -- (5,2)
      (2,0) -- (3,0) -- (3,5) -- (2,5)
      (1,1) -- (4,1) -- (4,4) -- (1,4);
\draw[red,->]
      (0,3) -- (5,3) -- (5,2) -- (0,2) -- (0,3);
\draw[blue,->]
      (3,0) -- (3,5) -- (2,5) -- (2,0) -- (3,0);
\draw[->]
      (1,1) -- (4,1) -- (4,4) -- (1,4) -- (1,1);
\end{tikzpicture}
```

Figure 5.8

Using the even odd rule. There are three rectangular sub-paths. For the even odd rule an area is filled if it requires the crossing of an odd number of lines to get from inside the area to "infinity."

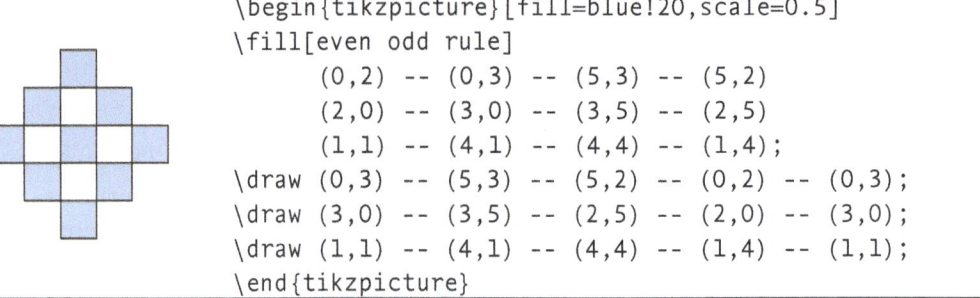

```
\begin{tikzpicture}[fill=blue!20,scale=0.5]
\fill[even odd rule]
      (0,2) -- (0,3) -- (5,3) -- (5,2)
      (2,0) -- (3,0) -- (3,5) -- (2,5)
      (1,1) -- (4,1) -- (4,4) -- (1,4);
\draw (0,3) -- (5,3) -- (5,2) -- (0,2) -- (0,3);
\draw (3,0) -- (3,5) -- (2,5) -- (2,0) -- (3,0);
\draw (1,1) -- (4,1) -- (4,4) -- (1,4) -- (1,1);
\end{tikzpicture}
```

point is considered inside the shape if a semi-infinite line originating at the point crosses an odd number of paths. Figure 5.8 depicts an example. This example also does not have self-overlapping paths.

5.9 Nodes and Node Labels

Diagrams with lines only are rare. Usually, they also contain text, math, or both. Fortunately, `tikz` has a mechanism for adding text, math, and other material to paths. This is done with the `node` path operation.

`\path ... node(⟨label⟩)[⟨options⟩]{⟨content⟩} ... ;`

The `node` path extension operation places ⟨content⟩ at the current position in the path using the options ⟨options⟩ and associates the label ⟨label⟩ with the node. The outer shape of the node is only drawn if `draw` is part of ⟨options⟩. The default shape is a rectangle but other shapes are also defined. The next section explains how to control the node shape. The texts (⟨label⟩) and [⟨options⟩] are optional. ☑

`\draw ... node(⟨label⟩)[⟨options⟩]{⟨content⟩} ... ;`

This is similar to the previous `\path` command. ☑

For example, the command `\draw (0,0) node {hello};` draws the word 'hello' at the origin. Likewise, the following draws a circle and the word 'circle' at position $(1, 0)$.

```
\draw (0,1)         % make (0,1) current position.   LaTeX Input
      circle (2pt)  % draw shape circle and
      node {circle}; % word circle at current position.
```

north west　north　north east
west○ hello ○east
south west　south　south east

```
\begin{tikzpicture}
    [every node/.style=scale=0.7]
\draw (0,0) node(hello)[scale=1.25] {hello};
\draw (hello.north)        circle (2pt)
        node[anchor=south]          {north};
\draw (hello.north east) circle (2pt)
        node[anchor=south west]     {north east};
... % remaining commands omitted.
```

Figure 5.9
Nodes and implicit labels

When a node receives a label, ⟨label⟩, then usually the additional labels ⟨label⟩.center, ⟨label⟩.north, ⟨label⟩.north east, ..., and ⟨label⟩.north west are also defined. The positions of these labels correspond to their names, so ⟨label⟩.north is to the north of the node having label ⟨label⟩. This holds for the most common node shapes. Figure 5.9 provides an example that involves all these auxiliary labels, except for ⟨label⟩.center. The option anchor in this example is explained further on. Basically, it provides a way to override the node's default insertion point.

5.9.1 Predefined Nodes Shapes

The previous section explained how to draw nodes. Nodes have a shape/style and content. The default node shape is rectangular but tikz also predefines the shapes coordinate, rectangle, circle, and ellipse. The remainder of this section presents some of the basic node shapes. The option shape=⟨shape⟩ determines the node shape. The following are the basic predefined node shapes.

coordinate This shape is for coordinates. Coordinates have no ⟨content⟩. They are not drawn but their positions are used as part of the picture.

rectangle This shape is for rectangular nodes. The rectangle is fitted around the ⟨content⟩. This is the default option.

circle This shape is for circles. The circle is fitted around the ⟨content⟩.

ellipse This shape is for ellipses. The ellipse is fitted around the ⟨content⟩.

The default height and width of a node may not always be ideal. Fortunately, there are options for low-level control. The minimum width, height, and size of a node are controlled with the options minimum width=⟨dimension⟩, minimum height=⟨dimension⟩, and minimum size =⟨dimension⟩. All these options work "as expected." There are also options for specifying the maximum width, height, and size of a node.

There are also options to set the *inner separation* and the *outer separation* of the node. Here the inner separation is the extra space between the bounding box of ⟨content⟩ and the node shape. For example, for a rectangular node, the inner separation determines the amount of space between the content of the node and its rectangle. Likewise, the outer separation is the extra space on the outside of the shape of the node. Both settings affect the size of the node and the

Figure 5.10

Low-level node control

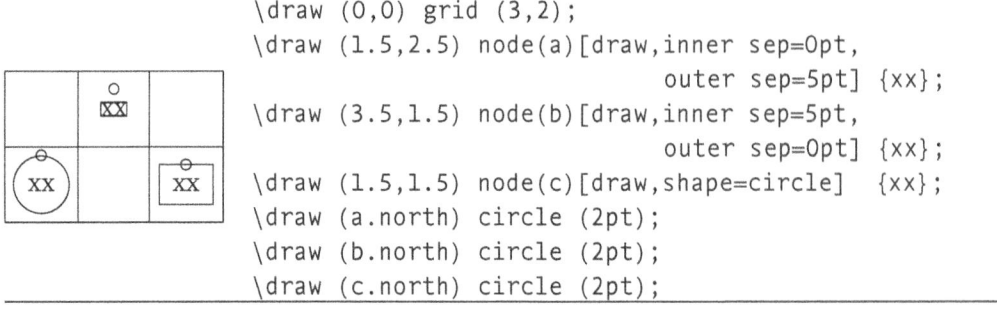

```
\draw (0,0) grid (3,2);
\draw (1.5,2.5) node(a)[draw,inner sep=0pt,
                               outer sep=5pt] {xx};
\draw (3.5,1.5) node(b)[draw,inner sep=5pt,
                               outer sep=0pt] {xx};
\draw (1.5,1.5) node(c)[draw,shape=circle]  {xx};
\draw (a.north) circle (2pt);
\draw (b.north) circle (2pt);
\draw (c.north) circle (2pt);
```

positions of the auxiliary labels north, north east, and so on. The options inner sep=⟨dimension⟩ and outer sep=⟨dimension⟩ set the inner and outer separation.

The options inner xsep=⟨dimension⟩, outer xsep=⟨dimension⟩, inner ysep=⟨dimension⟩, and outer ysep=⟨dimension⟩ control the separations in the horizontal and vertical directions. They work "as expected".

Figure 5.10 shows some of the different node shape options and low-level control. The difference in the inner separations of the rectangular nodes manifests itself in different sizes for the rectangular shapes. Differences in the outer separations result in different distances of labels such as north. The higher the outer separation of a node, the further its north label is away from its rectangular shape.

5.9.2 *Node Options*

This section briefly explains some node options that affect the drawing of nodes.

draw

This forces the drawing of the node shape as part of a \path command. By default the drawing of nodes is off. ☑

scale=⟨factor⟩

This scales the drawing of the node content by a factor of ⟨factor⟩. This includes the font size, line widths, and so on. ☑

anchor=⟨anchor⟩

This defines the *anchor* of the node, which is useful for positioning a node relative to a given point. This options draws the node such that the anchor coincides with the current position in the path. All node shapes define the anchor center, but most node shapes also define the compass directions north, north east, east, ..., and north west. The standard shapes also define base, base east, and base west. With the option base the insertion point is on the baseline of the node and the centre of the node is above the insertion point. The other options are directional versions. The options mid, mid east, and mid west are also defined for the standard nodes. With mid the insertion point is the mid point of the node—it is above the base point, between the baseline and the midline. The default value for ⟨anchor⟩ is center. ☑

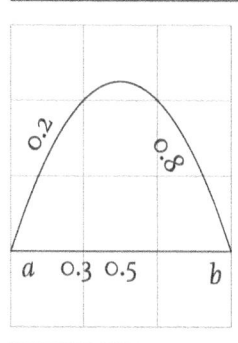

```
\draw[help lines] (0,0) grid (3,4);
\draw (0,1) coordinate(a)
         node[anchor=north west] {$a$}
      -- (3,1) coordinate(b)
         node[anchor=north east] {$b$}
         node[pos=0.3,anchor=north] {$0.3$}
         node[pos=0.5,anchor=north] {$0.5$}
     (a) .. controls (1,4) and (2,4) .. (b)
         node[pos=0.2,sloped,anchor=south] $0.2$
         node[pos=0.8,sloped,anchor=north] $0.8$;
```

Figure 5.11

Node placement

shift=⟨shift⟩

This option shifts the node in the direction ⟨shift⟩. Here ⟨shift⟩ is a regular coordinate or a label (including the parentheses). There are also directional versions xshift=⟨dimension⟩ and yshift=⟨dimension⟩ for horizontal and vertical shifting. ☑

above

This is equivalent to anchor=south. The options below, left, right, above left, above right, below left, and below right work in a similar way. ☑

above=⟨shift⟩

This combines the options anchor = south and shift=⟨shift⟩. The options left=⟨shift⟩, right=⟨shift⟩, ..., work in a similar way. ☑

rotate=⟨angle⟩

Draws the node, after rotating it ⟨angle⟩ degrees about its anchor. ☑

pos=⟨real⟩

This option is for placing nodes along a path (as opposed to at the current coordinate). This option places the node at the relative position on the path that is determined by ⟨real⟩, so if ⟨real⟩ is equal to 0.5 then the node is drawn mid-way, if it is equal to 1 then it is drawn at the end, and so on. ☑

pos=sloped

This very useful option rotates the node such that its base line is parallel to the tangent of the path where the node is drawn. ☑

midway

This option is equivalent to pos=0.5. Likewise, the option start is equivalent to pos=0, very near start is equivalent to pos=0.125, near start is equivalent to pos=0.25, near end is equivalent to pos=0.75, very near end is equivalent to pos=0.875, and end is equivalent to pos=1. ☑

Figure 5.11 shows some of these node options. Notice that several nodes can be placed with pos options for the same path segment.

5.9.3 Connecting Nodes

The tikz package is well-behaved. It won't cross lines unless you say so. This includes the crossing of borderlines of node shapes. For example, let's assume you've created two nodes. One of them is a circle, which is

Figure 5.12

Drawing lines between node shapes

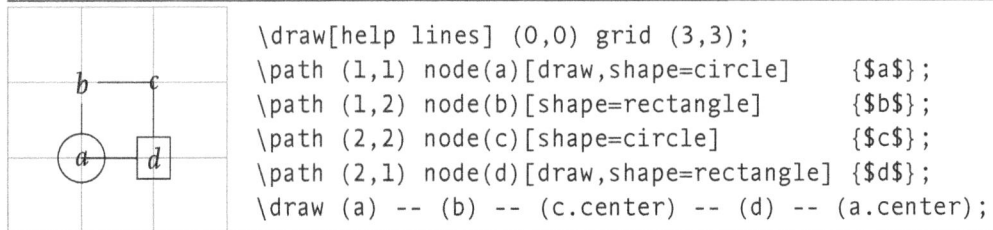

```
\draw[help lines] (0,0) grid (3,3);
\path (1,1) node(a)[draw,shape=circle]    {$a$};
\path (1,2) node(b)[shape=rectangle]      {$b$};
\path (2,2) node(c)[shape=circle]         {$c$};
\path (2,1) node(d)[draw,shape=rectangle] {$d$};
\draw (a) -- (b) -- (c.center) -- (d) -- (a.center);
```

Figure 5.13

The circle split node style

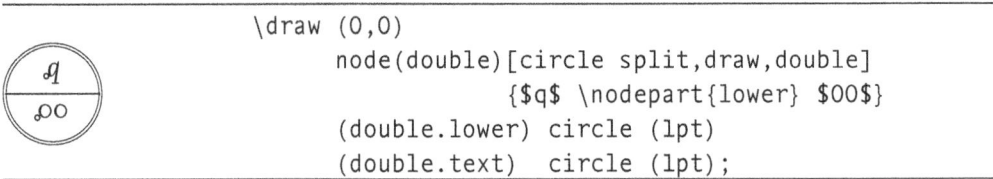

```
\draw (0,0)
        node(double)[circle split,draw,double]
                    {$q$ \nodepart{lower} $00$}
      (double.lower) circle (1pt)
      (double.text)  circle (1pt);
```

labelled ⟨c⟩, and the other is a rectangle, which is labelled ⟨r⟩. When you draw a line using the command \draw (⟨c⟩) -- (⟨r⟩); then the resulting line segment will *not* join the centres of the two nodes. The actual line segment will be shorter because the line segment starts at the circle shape and ends at the rectangle shape. In most cases this is the desired behaviour. If you need a line between the centres then you can always use the .center notation. Figure 5.12 provides an example.

5.9.4 *Special Node Shapes*

We've already seen that tikz has coordinate, circle, rectangle, and ellipse shape styles. Loading the tikz library shapes makes more shape styles available. You load this library by including it with the command \includetikzlibrary{shapes} in your document preamble. The following are some interesting shape styles.

circle split

This defines a circle with a text and a lower *node part*. The node parts of the circle split are separated by a horizontal line. The node part text is the upper and the node part lower is the lower part of the node shape circle split. The argument of the node shape determines what is in the node parts. This works as follows. You start by typesetting the default node part, which is called text. Next you may switch to a different node part and typeset that node part. This may be done several times. The command \nodepart{⟨part⟩} switches to the node part ⟨part⟩. After switching to the node part ⟨part⟩ you provide commands that typeset ⟨part⟩. For example, \node[shape=circle split]{top \nodepart{lower} bottom} typesets a circle split whose text part has 'top' in it and whose lower part has 'bottom' in it. The node shape circle split inherits all labels from the node shape circle. It also gets a label for the lower part. The node options circle split and double in Figure 5.13 result in a split circle with a double line. ☑

ellipse split

This is the ellipse version of circle split.

Figure 5.14

A node with rectangle style and several parts

Row 1	
Row 2	
	Row 3
Row four	

```
\node[rectangle split, rectangle split parts=4,
    every text  node part/.style={align=center},
    every two   node part/.style={align=left},
    every three node part/.style={align=right},
    draw, text width=2.5cm]
  { Row 1
    \nodepart{two} Row 2
    \nodepart{three} Row 3
    \nodepart{four} Row four };
```

```
\draw (0,0) node[ellipse split,draw]
            {hi \nodepart{lower} lo};  ☑
```

rectangle split

This is the rectangle version of `circle split`. The rectangle shape has many options and can split horizontally or vertically into up to 20 parts. There are quite a number options for this shape. The example in Figure 5.14 draws a rectangle with four parts. The example only works if the text width is set explicitly. The reader is referred to the tikz manual [Tantau 2010] for further information.   ☑

5.10 The spy Library

The spy library lets you magnify parts of diagrams. These magnifications are technically known as *canvas transformations*, which means they affect everything, including line widths, font size, and so on.

To use the feature you must add the option spy scope to the picture or scope you wish to spy upon. Some options implicitly add this option. Figure 5.15 shows an example of the spy feature. The notation $(\alpha{:}r)$ in the example is a polar coordinate. If is equivalent to the coordinate $(r\cos\alpha, r\sin\alpha)$. Polar coordinates are explained in further detail in Section 5.14. The spy command has many of options. If you like spying on your tikzpictures then you may find more details in the manual [Tantau 2010].

5.11 Trees

Knowing how to define node labels and knowing how to draw nodes and basic shapes, we are ready to draw some more interesting objects. We shall start with a class of objects that should be of interest to the majority of computer scientists: trees.

Trees expose a common theme in tikz objects: hierarchical structures. A tree is defined by defining its root and the children of each node in the tree. Each child is a node or a node with children. By default, the children of each parent are drawn from left to right in order of appearance. Unfortunately, drawing trees with tikz isn't per-

Figure 5.15

Using the spy library

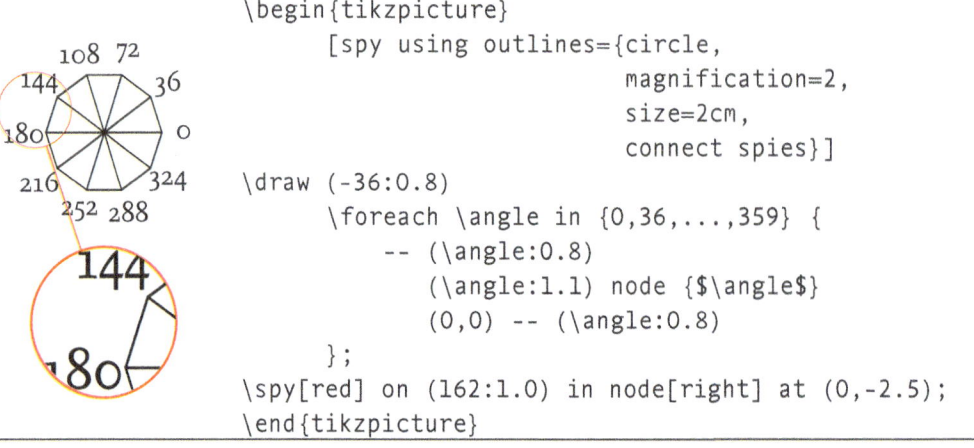

```
\begin{tikzpicture}
    [spy using outlines={circle,
                         magnification=2,
                         size=2cm,
                         connect spies}]
\draw (-36:0.8)
    \foreach \angle in {0,36,...,359} {
        -- (\angle:0.8)
            (\angle:1.1) node {$\angle$}
            (0,0) -- (\angle:0.8)
    };
\spy[red] on (162:1.0) in node[right] at (0,-2.5);
\end{tikzpicture}
```

Figure 5.16

Drawing a tree

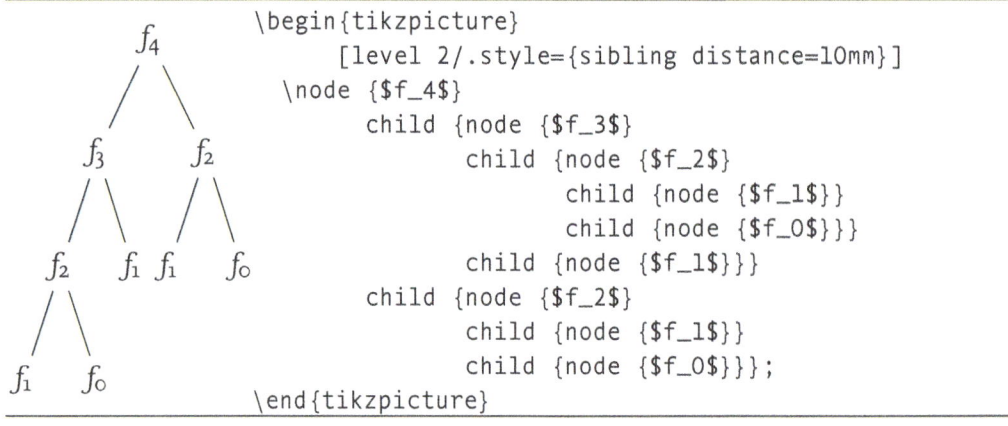

```
\begin{tikzpicture}
    [level 2/.style={sibling distance=10mm}]
    \node {$f_4$}
        child {node {$f_3$}
            child {node {$f_2$}
                child {node {$f_1$}}
                child {node {$f_0$}}}
            child {node {$f_1$}}}
        child {node {$f_2$}
            child {node {$f_1$}}
            child {node {$f_0$}}};
\end{tikzpicture}
```

Figure 5.17

Using implicit node labels in trees. To draw the arrow, the label of the root node is used to construct the labels of its first and second child.

```
\node (top) {$f_3$}
    child {node {$f_1$}}
    child {node {$f_2$}
        child {node {$f_1$}}
        child {node {$f_0$}}};
\draw[-angle 90]
    (top-1.north east) .. controls (top.south)
                        .. (top-2.north west);
```

fect. The `sibling distance=⟨dimension⟩` option lets you control the sibling distance. You can control these distances globally or for a fixed level. For example, `level 2/.style={sibling distance=1cm}` sets the distance for the grandchildren of the root—they are at level 2—to 1 cm. Figure 5.16 demonstrates how to draw a tree.

Inside trees labels work as usual. What is more, `tikz` implicitly labels the nodes in the tree. The *i*th child of a parent with label ⟨parent⟩ is labelled ⟨parent⟩-*i*. This process is continued recursively, so the *j*th child of the *i*th child of the parent node is labelled ⟨parent⟩-*i*-*j*. Figure 5.17 demonstrates the mechanism.

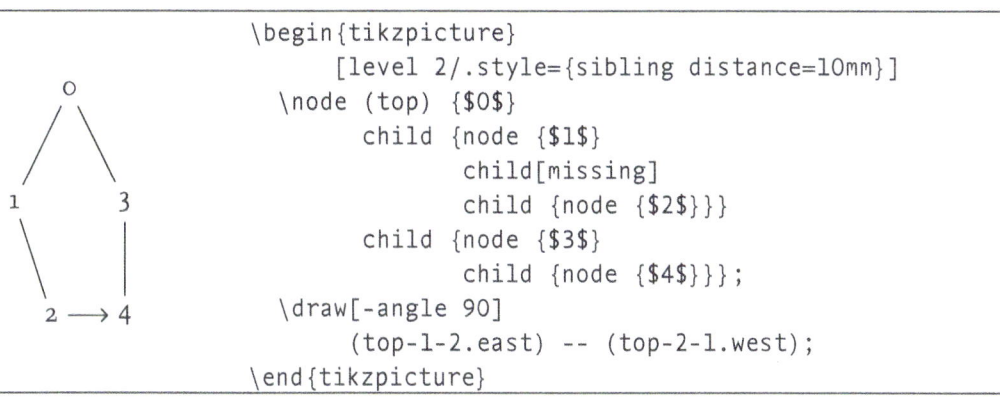

```
\begin{tikzpicture}
    [level distance=10mm%
    ,every node/.style={fill=red!60,%
                        circle,%
                        draw=black,%
                        inner sep=1pt}%
    ,level 1/.style={sibling distance=15mm},%
    ,level 2/.style={sibling distance=10mm,%
                     nodes={fill=red!20}}]
\node (top) {$f_3$}
    child {node[fill=blue!40] {$f_1$}}
    child {node[fill=blue!20] {$f_2$}
        child {node {$f_1$}}
        child {node {$f_0$}}};
\end{tikzpicture}
```

Figure 5.18
Controlling the node style

```
\begin{tikzpicture}
    [level 2/.style={sibling distance=10mm}]
\node (top) {$0$}
    child {node {$1$}
        child[missing]
        child {node {$2$}}}
    child {node {$3$}
        child {node {$4$}}};
\draw[-angle 90]
    (top-1-2.east) -- (top-2-1.west);
\end{tikzpicture}
```

Figure 5.19
A tree with a 'missing' node. The node of the first child of the root's first child is left out using the node option missing.

You can change the node style per tree level. Figure 5.18 uses different nodes styles for the second and third level. The option level distance=⟨dimension⟩ sets the distance between the levels in the tree.

As already noted, the rules for automatic node placement are not always ideal. For example, sometimes you may wish to have the single child of a given parent drawn to the left or to the right of the parent. The child option missing allows you to specify a node that takes up space but that is not drawn. By putting such a node to the left of its sibling, the position of the sibling is forced to the right. You may use this mechanism to force node placement. Omitting a node makes its label inaccessible. Figure 5.19 provides an example.

5.12 Logic Circuits

A *logic circuit* is a circuit whose building blocks are logic gates such as and-gates, or-gates, xor-gates (exclusive or), not-gates, and so on. Needless to say that tikz lets you draw logic circuits with ease. The style of the symbols of the gates depends on libraries. Possible libraries are circuits.logic.IEC, circuits.logic.CDH, and circuits.logic.US.

Table 5.4

Node shapes provided by logic gate shape libraries

Node Shape	Appearance			Node Shape	Appearance		
	IEC	CDH	US		IEC	CDH	US
and gate				nand gate			
or gate				nor gate			
xor gate				xnor gate			
buffer gate				not gate			

You may load these libraries with the \usetikzlibrary command. Table 5.4 shows the node shapes for the different libraries. The options used to draw the shapes are given by {circuit, logic ⟨style⟩, tiny circuit symbols, every circuit symbol/.style={fill=white, draw}}, where ⟨style⟩ is IEC, CDH, or US. The option logic gate IEC symbol color=black was also used when IEC was used.

Figure 5.20 draws a logic circuit with tikz. There are two new concepts in this example. The first is the option circuit declare symbol, which defines names of new circuit symbols. The second new concept is set ⟨symbol⟩ graphic, which defines the appearance of the circuit symbol ⟨symbol⟩. These options define two new symbols called connection and io. The former draws connections as black filled circles. The latter draws the input and output nodes as circles. The style of the remaining symbols is determined by the circuit logic CDH option. Notice that this example requires the library circuits.logic.CDH. Finally, note that if you're drawing many circuits you can define the defaults for your circuits with the \tikzset command.

5.13 Commutative Diagrams

Commutative diagrams are frequently used in mathematics and theoretical computer science. The purpose of these diagrams is to provide some high-level overview of the composition of functions/morphisms.

The input in Figure 5.21 uses the \matrix command to construct a commutative diagram for the Homomorphism Theorem. The output is presented in Figure 5.22. In this example, which is based on [Becker, and Weispfenning 1993, Theorem 1.55], $\phi\colon R \to S$ is a homomorphism of rings, I is an ideal of R with $I \subseteq \ker(\phi)$, χ is the canonical homomorphism from R to R/I, and $\psi\colon R/I \to S$ is the ring homomorphisms satisfying $\psi \circ \chi = \phi$. Note that the nodes in the example contain mathematical content of the form $⟨stuff⟩$. This is explained in Chapter 8.

The example uses the \matrix command to define a matrix of rows with alignment positions. As is usual, a row is ended with a double backslash and columns are separated with ampersand characters. Note that it is required to explicitly terminate the last row.

Figure 5.20
Drawing a half adder with tikz

```
\begin{tikzpicture}
    [circuit logic CDH,
     circuit declare symbol=connection,
     circuit declare symbol=io,
     set connection graphic={fill=black,
                             shape=circle,
                             minimum size=1mm},
     set io graphic={draw,shape=circle,
                     minimum size=1mm},
     every circuit symbol/.style={fill=white,draw}]
\draw                       node[xor gate] (x)  {}
  +(0,-1)                   node[and gate] (a)  {}
  (x.input 1) +(-0.8,0)     node[io]       (A)  {}
  (A |- a.input 2)          node[io]       (B)  {}
  (x.output) -- +(0.4,0)    node[io]       (S)  {}
  (a.output) -- (a.output -| S)
                            node[io]       (C)  {}
  ($(B)!0.33!(a.input 2)$)  node[connection]    {}
    |- (x.input 2)
  ($(A)!0.66!(x.input 1)$)  node[connection]    {}
    |- (a.input 1)
  (A.west)                  node[anchor=east]   {$A$}
  (B.west)                  node[anchor=east]   {$B$}
  (S.east)                  node[anchor=west]   {$S$}
  (C.east)                  node[anchor=west]   {$C$}
  (A) -- (x.input 1)
  (B) -- (a.input 2);
\end{tikzpicture}
```

Inside the matrix, you can use tikz commands such as \draw. You also can define labels. In the example, there is only one command per matrix cell, but you may have more commands.

Note that the symbols near the arrows should be near the centres of these arrows. We could have placed these nodes explicitly using a node command of the form \node[midway]{⟨stuff⟩}, but that would mean a lot of repetition. Instead we define the style of the nodes to be midway in the optional argument of the tikzpicture environment.

The distance between the rows and columns is defined in terms of ems. That way, the diagram scales with commands such as \large. Using centimetres or other absolute distances is not such a good idea.

Finally, the example doesn't really need an explicit placement of the matrix (using at). The placement is only added to show that matrices can be placed at a certain position.

5.14 Coordinate Systems

Specifying coordinates is the key to effective, efficient, and maintainable picture creation. Coordinates may be specified in different ways

Figure 5.21

Input of commutative diagram. Most of the construction should be clear. Setting the node style to `midway` in the options of the `tikzpicture` forces the three drawn nodes at the bottom between the start and end points of the arrows.

```
\begin{tikzpicture}[every node/.style={midway}]
\matrix[column sep={4em,between origins},
        row sep={2em}] at (0,0) {
    \node(R)    {$R$};   & \node(S)    \\
    \node(R/I)  {$R/I$}; \\ % don't omit previous \\
};
\draw[->] (R)   -- (R/I) node[anchor=east]  {$\chi$};
\draw[->] (R/I) -- (S)   node[anchor=north] {$\psi$};
\draw[->] (R)   -- (S)   node[anchor=south] {$\phi$};
\end{tikzpicture}
```

Figure 5.22

Commutative diagram. This diagram is the result of the input in Figure 5.21.

each coming with its own specific *coordinate system*. Within a coordinate system you specify coordinates using *explicit* or *implicit* notation.

explicit Explicit coordinate specifications are verbose. To specify a coordinate, you write (⟨system⟩ cs:⟨coord⟩), where ⟨system⟩ is the name of the coordinate system and where ⟨coord⟩ is a coordinate whose syntax depends on ⟨system⟩. For example, to specify the point having *x*-coordinate ⟨x⟩ and *y*-coordinate ⟨y⟩ in the canvas coordinate system you write (canvas cs:x=⟨x⟩,x=⟨y⟩).

implicit Implicit coordinates specifications are shorter than explicit coordinate specifications. You specify coordinates using some coordinate system-specific notation inside parentheses. Most examples so far have used the implicit notation for the canvas coordinate system.

The remainder of this section studies some of the more useful coordinates systems. The notation for explicit coordinate specification being too verbose, we shall focus on using implicit notation.

canvas coordinate system The most widely used coordinate system is the canvas coordinate system. It defines coordinates in terms of a horizontal and a vertical offset relative to the origin. The implicit notation (⟨x⟩,⟨y⟩) is the point with *x*-coordinate ⟨x⟩ and *y*-coordinate ⟨y⟩.

xyz coordinate system The xyz coordinate system defines coordinates in terms of a linear combination of an *x*-, a *y*-, and a *z*-vector. By default, the *x*-vector points 1 cm to the right, the *y*-vector points 1 cm up, and the *z*-vector points to $(-\sqrt{2}/2, -\sqrt{2}/2)$. However, these default settings can be changed. The implicit notation (⟨x⟩,⟨y⟩,⟨z⟩) is used to define the point at ⟨x⟩ times the *x*-vector plus ⟨y⟩ times the *y*-vector plus ⟨z⟩ times the *z*-vector.

polar coordinate system The canvas polar coordinate system defines coor-

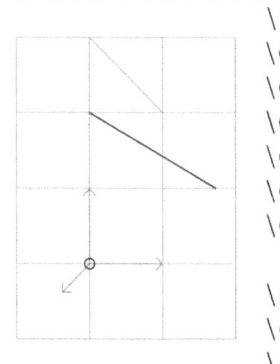

```
\begin{tikzpicture}[>=angle 90]
\draw[help lines] (-1,-1) grid (2,3);
\draw[red] (canvas cs:x=1cm,y=2cm) -- (0,3);
\draw[blue,->] (0,0) -- (xyz cs:x=1,y=0,z=0);
\draw[blue,->] (0,0) -- (0,1,0);
\draw[blue,->] (0,0) -- (0,0,1);
\draw (canvas polar cs:radius=2cm,angle=30)
        -- (90:2);
\path (0,0) coordinate (origin);
\draw (origin) node circle (2pt);
\end{tikzpicture}
```

Figure 5.23
Using four coordinate systems

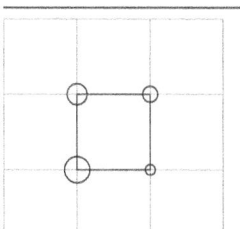

```
\draw[help lines] (0,0) grid +(3,3);
\path (1,1) coordinate (ll);
\path (2,2) coordinate (ur);
\draw (ll) -- (ll -| ur) circle (2pt);
\draw (ll -| ur) -- (ur) circle (3pt);
\draw (ur) -- (ur -| ll) circle (4pt);
\draw (ur -| ll) -- (ll) circle (5pt);
```

Figure 5.24
Computing the intersection of
perpendicular lines

dinates in terms of an angle and a radius. The implicit notation $(\alpha\!:\!r)$ corresponds to the point $r \times (\cos\alpha, \sin\alpha)$. Angles in this coordinate system, as all angles in tikz, should be supplied in degrees.

node coordinate system The node coordinate system defines coordinates in terms of a label of a node or coordinate. The implicit notation $(\langle \text{label} \rangle)$ is the position of the node or coordinate that was given the label $\langle \text{label} \rangle$.

Figure 5.23 demonstrates the previous four coordinate systems in action. The optional argument of the tikzpicture sets the arrow head style to the predefined style angle 90.

perpendicular coordinate system The perpendicular coordinate system is a dedicated system for computing intersections of horizontal and vertical lines. With this coordinate system's implicit syntax you write $(\langle \text{pos}_1 \rangle$ $|\text{-}\ \langle \text{pos}_2 \rangle)$ for the coordinate at the intersection of the infinite vertical line though $\langle \text{pos}_1 \rangle$ and the infinite horizontal line through $\langle \text{pos}_2 \rangle$. Likewise, $(\langle \text{pos}_1 \rangle\ \text{-}|\ \langle \text{pos}_2 \rangle)$ is the intersection of the infinite horizontal line though $\langle \text{pos}_1 \rangle$ and the infinite vertical line through $\langle \text{pos}_2 \rangle$. The notation for this coordinate system is quite suggestive because | suggests vertical and - suggests horizontal. The order of the lines is then given by the order inside the operators |- and -|. Inside the parentheses you are not supposed to use parentheses for coordinates and labels, so you write (0,1 |- 1,2), (label |- 1,2), and so on. Figure 5.24 demonstrates how to use the perpendicular coordinate system.

You can freely mix the coordinate systems. For example \draw (0,0) -- (0,1); and \draw (0,0) -- (90:1); are equivalent.

Figure 5.25

Absolute, relative, and incremental coordinates

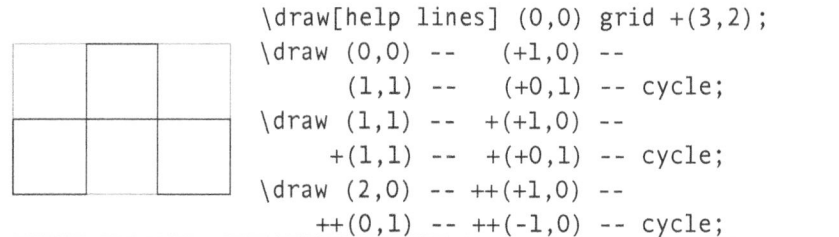

```
\draw[help lines] (0,0) grid +(3,2);
\draw (0,0)  --    (+1,0) --
          (1,1) --    (+0,1) -- cycle;
\draw (1,1) --  +(+1,0) --
       +(1,1) --  +(+0,1) -- cycle;
\draw (2,0) -- ++(+1,0) --
      ++(0,1) -- ++(-1,0) -- cycle;
```

5.15 Coordinate Calculations

Specifying diagrams in terms of absolute coordinates is cumbersome and prone to errors. What is worse, diagrams defined in terms of absolute coordinates are difficult to maintain. For example, changing the position of an n-agon that is defined in terms of absolute coordinates requires changing n coordinates. Fortunately, tikz lets you compute coordinates from other coordinates. Used intelligently, this reduces the maintenance costs of your diagrams.

There are two kinds of coordinate computations. The first kind involves *relative* and *incremental* coordinates. These computations depend on the current coordinate in a path. They are explained in Section 5.15.1. The second kind of computations is more general. They can be used to compute coordinates from coordinates, distances, rotation angles, and projections. These computations are explained in Section 5.15.2.

5.15.1 Relative and Incremental Coordinates

Relative and *incremental* coordinates are computed from the current coordinate in a path. The first doesn't change the current coordinate whereas the second does change it.

relative coordinate A relative coordinate constructs a new coordinate at an offset from the current coordinate without changing the current coordinate. The notation +⟨offset⟩ specifies the relative coordinate that is located at offset ⟨offset⟩ from the current coordinate.

incremental coordinate An incremental coordinate also constructs a new coordinate at an offset from the current coordinate. This time, however, the new coordinate becomes the current coordinate. You use the implicit notation ++⟨offset⟩ for incremental coordinates.

Figure 5.25 provides an example that draws three squares. The first square is drawn with absolute coordinates, the second with relative coordinates, and the last with incremental coordinates. Clearly, the relative and incremental coordinates should be preferred because they improve the maintenance of the picture. For example, moving the first square requires changing four coordinates, whereas moving the second or third square requires changing only the start coordinate. The relative coordinate in the grid also improves the maintainability.

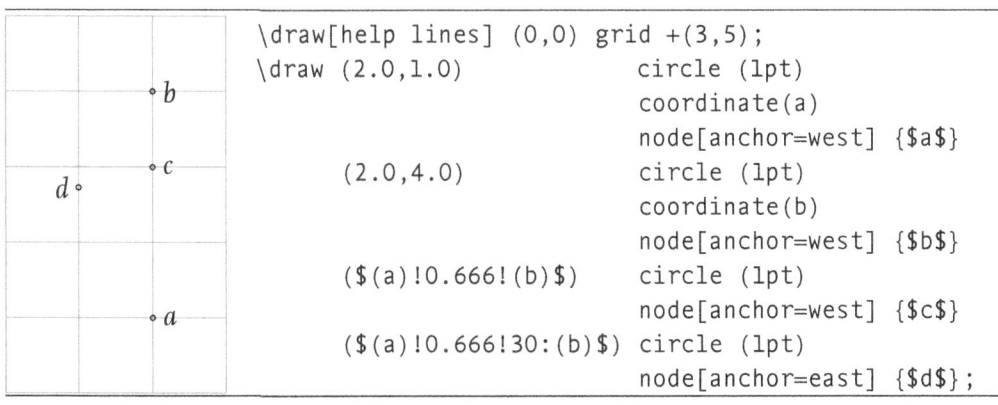

```
\draw[help lines] (0,0) grid +(3,5);
\draw (2.0,1.0)            circle (1pt)
                           coordinate(a)
                           node[anchor=west] {$a$}
      (2.0,4.0)            circle (1pt)
                           coordinate(b)
                           node[anchor=west] {$b$}
      ($(a)!0.666!(b)$)    circle (1pt)
                           node[anchor=west] {$c$}
      ($(a)!0.666!30:(b)$) circle (1pt)
                           node[anchor=east] {$d$};
```

Figure 5.26

Computations with partway modifiers

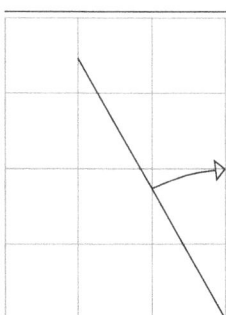

```
\draw[help lines] (-3,0) grid +(3,4);
\draw (0,0) --
      ($(0,0)!  1! 30:(0,4)$) coordinate(a)
      ($(0,0)!2cm!    (a)$)   coordinate(b)
      ($(0,0)!2cm!-15:(a)$)   coordinate(c)
      ($(0,0)!2cm!-30:(a)$)   coordinate(d);
\draw[-open triangle 90]
      (b) .. controls (c) .. (d);
```

Figure 5.27

Computations with partway and distance modifiers

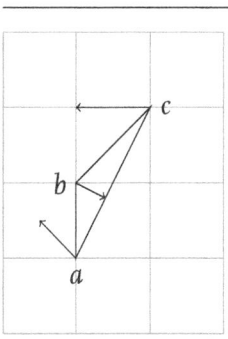

```
\begin{tikzpicture}[>=open triangle 90]
\draw[help lines] (0,0) grid +(3,4);
\draw    (1,1) coordinate(a) node[anchor=north] {$a$}
      -- (1,2) coordinate(b) node[anchor=east]  {$b$}
      -- (2,3) coordinate(c) node[anchor=west]  {$c$}
      -- cycle;
\draw[->]  (b) -- ($(a)!(b)!(c)$);
\draw[->]  (c) -- ($(b)!(c)!(a)$);
\draw[->]  (a) -- ($(c)!(a)!(b)$);
\end{tikzpicture}
```

Figure 5.28

Computations with projection modifiers

5.15.2 *Complex Coordinate Calculations*

Finally, tikz offers complex coordinate calculations. However, these calculations are only available if the tikz library calc is loaded in the preamble: \usetikzlibrary{calc}.

([⟨options⟩]$⟨coordinate computation⟩$)

This is the general syntax. The ⟨coordinate computation⟩ should:

1. Start with ⟨factor⟩*⟨coordinate⟩⟨modifiers⟩. Here ⟨modifiers⟩ is a sequence of one or more ⟨modifier⟩s and ⟨factor⟩* is an optional multiplication factor, which defaults to 1. Both are described further on.

2. Continue with one or more expressions of the form: ⟨sign option⟩

⟨factor⟩*⟨coordinate⟩⟨modifiers⟩, where ⟨sign option⟩ is an optional + or -. ☑

⟨factor⟩

Each ⟨factor⟩ is an optional numeric expression that is parsed by the \pgfmathparse command. Examples of valid ⟨factor⟩s are 1.2, {3 * 4}, {3 * sin(60)}, {3 + (2 * 4)}, and so on. Inside the braces it is safe to use parentheses, except for the top level. The reason why parentheses do not work at the top level is that ⟨factor⟩s are optional and that the opening parenthesis is reserved for the start of a coordinate. Therefore, compound expressions at the top level are best put inside braces as this makes parsing easier at the top level. ☑

⟨modifier⟩

A ⟨modifier⟩ is a postfix operator acting on the coordinate preceding it. There are three different kinds: ⟨pmod⟩, ⟨dmod⟩, and ⟨prmod⟩. Each of them is of the form !⟨stuff⟩ and it is used after a coordinate. To explain the modifiers we shall write ⟨partway modifier⟩ for ⟨coordinate⟩!⟨pmod⟩, shall write ⟨distance modifier⟩ for ⟨coordinate⟩!⟨dmod⟩, and shall write ⟨projection modifier⟩ for ⟨coordinate⟩!⟨prmod⟩. ☑

⟨partway modifier⟩

There are two forms of ⟨partway modifier⟩s. The first is ⟨coordinate$_1$⟩!⟨factor⟩!⟨coordinate$_2$⟩ The resulting coordinate is given by

$$\langle coordinate_2 \rangle + \langle factor \rangle \times (\langle coordinate_2 \rangle - \langle coordinate_1 \rangle).$$

In words this is the coordinate that is at ⟨factor⟩ × 100% distance along the line between ⟨coordinate$_1$⟩ and ⟨coordinate$_2$⟩.

A ⟨partway modifier⟩ may also have the complex form ⟨coordinate$_1$⟩!⟨factor⟩!⟨angle⟩:⟨coordinate$_2$⟩. This first computes ⟨coordinate$_1$⟩!⟨factor⟩!⟨coordinate$_2$⟩ and then rotates the resulting coordinate about ⟨coordinate$_1$⟩ over ⟨angle⟩ degrees. Figure 5.26 shows some coordinate computations with partway modifiers. ☑

⟨distance modifier⟩

The next modifier is the ⟨distance modifier⟩. This modifier has the form ⟨coordinate$_1$⟩!⟨distance⟩:⟨angle⟩!⟨coordinate$_2$⟩, where :⟨angle⟩ is optional.

The simpler form ⟨coordinate$_1$⟩!⟨distance⟩!⟨coordinate$_2$⟩ results in the coordinate that is at distance ⟨distance⟩ from ⟨coordinate$_1$⟩ in the direction from ⟨coordinate$_1$⟩ to ⟨coordinate$_2$⟩. For example, if the two coordinates are 2 cm apart then setting ⟨distance⟩ to 1cm gives you the point halfway between the two coordinates.

The more complex form of the distance modifier is similar to and works in a similar way as the partway modifier. This time you write ⟨coordinate$_1$⟩!⟨distance⟩:⟨angle⟩!⟨coordinate$_2$⟩. The result is obtained by first computing ⟨coordinate$_1$⟩!⟨distance⟩!⟨coordinate$_2$⟩ and rotating the result about ⟨coordinate$_1$⟩ over ⟨angle⟩ degrees in a counter-clockwise direction. Figure 5.27 presents an example of coordinate computations involving distance modifiers. ☑

⟨projection modifier⟩

The final ⟨modifier⟩ is a ⟨projection modifier⟩. This modifier is of the form ⟨coordinate$_1$⟩!⟨coordinate$_2$⟩!⟨coordinate$_3$⟩[1] and it results in the projection of ⟨coordinate$_2$⟩ on the infinite line through ⟨coordinate$_1$⟩ and ⟨coordinate$_3$⟩. Figure 5.28 presents an example of coordinate computations with projection modifiers. (For some reason my tikz version doesn't like extra space around ! inside a ⟨projection modifier⟩. It is not clear whether this is a feature.) ☑

5.16 Options

Many tikz commands and environments depend on options. Usually these options are specified using ⟨key⟩=⟨value⟩ combinations. Some combinations have shorthand notations. For example, ⟨colour⟩ is a shorthand notation for color=⟨colour⟩. Options are best defined by passing their ⟨key⟩ = ⟨value⟩ combinations as part of the optional argument. However, there is another mechanism.

\tikzset{⟨options⟩}

Sets the options in ⟨options⟩. The options are set using the pgfkeys package. This package is quite powerful but explaining it goes beyond the scope of an introduction like this chapter. Roughly speaking, processing the keys works "as expected" for "normal" usage. There are no examples of how you redefinine existing options. However, the following section provides examples of how the \tikzset command may be used to set user-defined styles. ☑

5.17 Styles

One of the great features of tikz is *styles*. Defining a style for your graphics has several advantages.

control Styles control the appearance. By carefully designing a style for drawing auxiliary lines, you can draw them in a style that makes them appear less prominently in the picture. Other styles may be used to draw lines that should stand out and draw attention.

consistency Drawing and colouring sub-diagrams with a carefully chosen style guarantees a consistent appearance of your diagrams. For example, if you consistently draw help lines in a dedicated, easily recognisable style then it makes it easier to recognise them.

reusability Styles that are defined once can be reused several times.

simplicity Changing the appearance of a graphical element with styles with well-understood interfaces is much easier and leads to fewer errors.

refinement You can stepwise refine the way certain graphics are drawn. This lets you postpone certain design decisions while still letting you draw your diagrams in terms of the style. By refining the style at a later stage, you can fine-tune the drawing of all the relevant graphics.

---

[1]The manual on page 119 also mentions an angle but it is not explained how to use it....

Figure 5.29

Predefining options with the
\tikzset command

```
\tikzset{thick dashed/.style={thick,dashed}}
\begin{tikzpicture}
        [{help lines/.style={ultra thin,blue!30}]
\draw[thick dashed] (0,0) rectangle (1,1);
\draw[help lines]   (1,1) rectangle (2,2);
\end{tikzpicture}
```

maintenance This is related to the previous item. Unforeseen changes in global requirements can be implemented by making a few local changes.

Styles affect options. For example, the predefined help lines style sets draw to black!50 and sets line width to very thin. You can set a style at a global or at a local level. The following command defines a style at a global level.

\tikzset{⟨style name⟩/.style={⟨list⟩}}

This defines a new style ⟨style name⟩ and gives it the value ⟨list⟩, where ⟨list⟩ is a list defining the style. Once the style is defined it can be used and using it is equivalent to executing the items in ⟨list⟩. In its basic usage ⟨list⟩ is a list of ground options, but it is also possible to define styles that take arguments. This is explained in Chapter 13. The following example defines a style Cork, which uses thick, dashed lines.

```
\tikzset{Cork/.style={dashed,thick}}
\draw[Cork] (0,0) rectangle (1,1);                ☑
```

⟨style name⟩/.style={⟨list⟩}

This mechanism may be used to temporarily override the existing definition of ⟨style name⟩. Figure 5.29 provides an example.   ☑

5.18 Scopes

Scopes in tikzpicture environments serve a similar purpose as blocks in a programming language and groups in LATEX. They let you temporarily change certain settings inside the scope and restore the original settings when leaving the scope. In addition tikz scopes let you execute code at the start and end of a scope. Scopes in tikz are implemented as an environment called scope. Scopes depend on the following style.

every scope

This style is installed at the start of every scope. The style is empty initially. There are two techniques to define the style. The first technique is to use \tikzset{every scope/.style={⟨list⟩}}. The second technique sets the style using the option of a tikzpicture environment.

```
\begin{tikzpicture}[every scope/.style={⟨list⟩}]                LATEX Usage
...
\end{tikzpicture}                                               ☑
```

Figure 5.30
Using scopes

```
\begin{scope}[fill=gray!50]
   \fill (0.5,1.5) circle (0.5);
   \begin{scope}[fill=gray]
      \fill (1.5,0.5) circle (0.5);
   \end{scope}
   \fill (2.5,1.5) circle (0.5);
\end{scope}
\draw (0,0) rectangle (1,1)
        { [rounded corners]
           rectangle (2,2) }
        (2,1) rectangle (3,0);
```

Figure 5.31
The \foreach command

```
\foreach \pos/\text in
            {{0,0}/1,{1,0}/2,{1,1}/3,{0,1}/4}
   {
      \draw (\pos) node {\text};
   }
```

The following options execute code at the start and end of the scope.

execute at begin scope=⟨code⟩

This option executes ⟨code⟩ at the start of the scope. ☑

execute at end scope=⟨code⟩

This option executes ⟨code⟩ at the end of the scope. ☑

The tikz library scopes defines a shorthand notation for scopes. It lets you write { [⟨options⟩] ⟨stuff⟩ } for \begin{scope}[⟨options⟩] ⟨stuff⟩ \end{scope}. Interestingly you can also have scopes *inside* paths. However, options of a local scope in a path do not affect path options that apply to a whole path. For example, line thickness, colour and so on. Figure 5.30 depicts an example.

5.19 The \foreach Command

As if tikz productivity isn't enough, its pgffor library provides a very flexible foreach command.

\foreach ⟨macros⟩ in {⟨list⟩} {⟨statements⟩}

Here ⟨macros⟩ is a forward slash-delimited list of macros and ⟨list⟩ is a comma-delimited list of lists with forward slash-delimited values. For each list of values in ⟨list⟩, the \foreach command binds the *i*-th value of the list to the *i*-th macro in ⟨macros⟩ and then carries out ⟨statements⟩. Figure 5.31 shows an example. This example also shows that grouping may be used to construct values in ⟨list⟩ with commas. In general this is a useful technique. Since coordinates are very common, there is no need to turn coordinates into a group.

It is also possible to use \foreach inside the \path command. The

Table 5.5

Shorthand notation for the \foreach command. The notation in the upper part of the table involves ranges that depend on an initial value and a next value that determines the increment. The shorthand notation in the middle part depends only on the initial value and the final value in the range. Here the increment is 1 if the final value is greater than the initial value. Otherwise the increment is –1. The lower part of the table demonstrates the \foreach command also allows pattern-matching.

Command	Yields
`\foreach \x in {1,2,...,6} {\x,}`	1, 2, 3, 4, 5, 6,
`\foreach \x in {1,3,...,10} {\x,}`	1, 3, 5, 7, 9,
`\foreach \x in {1,3,...,11} {\x,}`	1, 3, 5, 7, 9, 11,
`\foreach \x in {0,0.1,...,0.3} {\x,}`	0, 0.1, 0.20001, 0.30002,
`\foreach \x in {a,b,...,d,9,8,...,6} {\x,}`	a, b, c, d, 9, 8, 7, 6,
`\foreach \x in {7,5,...,0} {\x,}`	7, 5, 3, 1,
`\foreach \x in {Z,X,...,M} {\x,}`	Z, X, V, T, R, P, N,
`\foreach \x in {1,...,5} {\x,}`	1, 2, 3, 4, 5,
`\foreach \x in {5,...,1} {\x,}`	5, 4, 3, 2, 1,
`\foreach \x in {a,...,e} {\x,}`	a, b, c, d, e,
`\foreach \x in {2^1,2^...,2^6} {$\x$,}`	$2^1, 2^2, 2^3, 2^4, 2^5, 2^6$
`\foreach \x in {0\pi,0.5\pi,...\pi,2\pi} {$\x$,}`	0π, 0.5π, 1.5π, 2.0π,
`\foreach \x in {A_1,..._1,D_1} {$\x$,}`	$A_1, B_1, C_1, D_1,$

following is an example, which also demonstrates that tikz supports a limited form of arithmetic.

```
\draw (0,-0.8)
\foreach \angle in {0,90,180,270} {
    -- (\angle:0.8)
        (\angle:1.0) + (\angle:0.2)
    \foreach \fraction in {1,2,3,4,5} {
        -- +(\angle+\fraction*72:0.2)
    } -- cycle (\angle:0.8)
};
```

If there is only one macro in ⟨macros⟩ then shorthand notations are allowed in ⟨list⟩. Some examples of these shorthand notations are listed in Table 5.5. The table is based on the tikz documentation.

5.20 The let Operation

The let operation binds expressions to "variables" inside a path. The following is the general syntax.

\path ... let ⟨assignments⟩ **in ...;**

Here ⟨assignments⟩ is a comma-delimited list of assignments. Each assignment is of the form ⟨register⟩ = ⟨expression⟩. To carry out the assignment, ⟨expression⟩ is evaluated and then assigned to ⟨register⟩, which is some local tikz variable. The macros \n, \p, \x, and \y can be used to get the defined values. These macros only work in the scope of the assignment, which lasts from the keyword in to the semicolon at the end of the path operation. The assignment mechanism respects the tikz scoping rules.

There are two kinds of ⟨register⟩ in assignments of the let operation. Both are written as macro calls. The first kind is number registers, which are written as \n{⟨name⟩}. Here ⟨name⟩ is just a convenient label, which may be almost any combination of characters, digits, space characters, and other regular characters. You cannot use the dot. As the name suggests, number registers store numeric values. The second kind of register is the point register. Point registers are written \p{⟨name⟩}. As the name suggests, they store coordinate values. The following explains both ⟨register⟩s in more detail.

\n{⟨name⟩} = ⟨expression⟩

This is for assigning a numeric value to the number register ⟨name⟩. The command \n{⟨name⟩} returns the current value of the number register ⟨name⟩. The following shows how to use number registers.

```
\draw let \n0 = 0, \n1 = 1 in
       let \n{the sum} = \n0 + \n1 in
       (0,0) -- (\n1,\n0) --
       (\n1,\n{the sum});
```

\p{⟨name⟩} = ⟨expression⟩

This is for assigning points to the point register ⟨name⟩. The command \p{⟨name⟩} returns the current (point) value of the point register ⟨name⟩. The x- and y-coordinates of the point register may be got with the commands \x{⟨name⟩} and \y{⟨name⟩}. The following is an example that assumes that the tikz library calc has been loaded. Note that this example can also be written with a single let operation.

```
\draw
let \p{ll} = (0,0),
    \p{ur} = (1,1) in
let \p{ul} = (\x{ll},\y{ur}),
    \p{lr} = ($\p{ll}!1!90:\p{ul}$) in
(\p{ll}) -- (\p{lr}) -- (\p{ur}) --
(\p{ul}) -- cycle;
```

5.21 The To Path Operation

The to path operation connects two nodes in a given style. For example, \path (0,0) to (1,0); is the same as \path (0,0) -- (1,0);. However, \path (0,0) to[out=45,in=135] (1,0); connects the points with an arc leaving (0, 0) at 45° and entering (1, 0) at 135°.

It is also possible to define styles for to operations. This lets you draw complex paths with a single operation. The general syntax of the to path operation is as follows.

\path ... to[⟨options⟩] ⟨nodes⟩ (⟨coordinate⟩) ...;

The ⟨nodes⟩ are optional nodes that are placed on the path. Figure 5.32 presents an example.

every to

This style is for to paths. The style is installed at the start of every to

Figure 5.32

Simple to path example

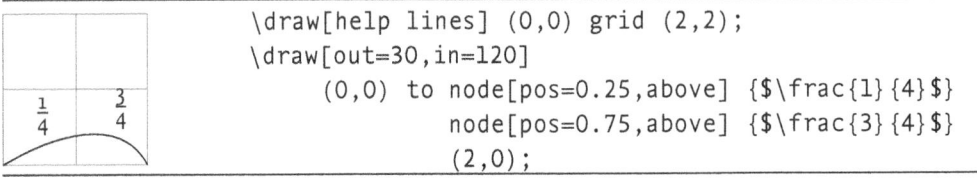

```
\draw[help lines] (0,0) grid (2,2);
\draw[out=30,in=120]
        (0,0)  to node[pos=0.25,above]  {$\frac{1}{4}$}
                  node[pos=0.75,above]  {$\frac{3}{4}$}
        (2,0);
```

Figure 5.33

user-defined to path. The \tikzset command defines a new to path style called hvh. Lines 2–5 are the style's actions. The command \tikztostart is the start of the path, \tikztotarget the destination, and \tikztonodes optional nodes at the end of the path. All these commands are determined by the \path command.

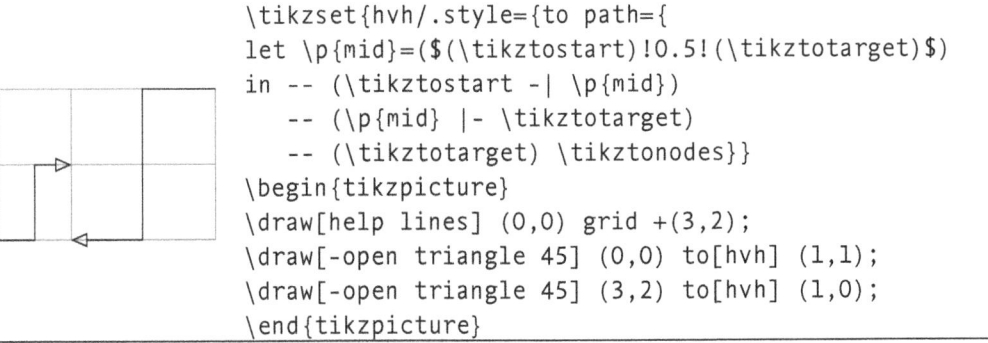

```
\tikzset{hvh/.style={to path={
let \p{mid}=($(\tikztostart)!0.5!(\tikztotarget)$)
in -- (\tikztostart -| \p{mid})
   -- (\p{mid} |- \tikztotarget)
   -- (\tikztotarget) \tikztonodes}}}
\begin{tikzpicture}
\draw[help lines] (0,0) grid +(3,2);
\draw[-open triangle 45] (0,0) to[hvh] (1,1);
\draw[-open triangle 45] (3,2) to[hvh] (1,0);
\end{tikzpicture}
```

path. The following shows the main ideas. We start by defining the style of the to path.

```
\tikzset{every to/.style={out=90,in=180}}
```
LaTeX Input

Of course you may also set the style by passing it as an option to a tikzpicture or a scope. Once the style is set, it becomes effective and is added to every to path.

```
\tikzset{every to/.style={out=90,in=180}}
\begin{tikzpicture}
\draw[help lines] (0,0) grid (2,2);
\draw (0,0) to (2,2)
      (0,0) to (1,1);
\end{tikzpicture}
```
☑

Options affect the style of to paths. The following is arguably the most important style.

to path=⟨path⟩

This inserts the path {[every to,⟨options⟩] ⟨path⟩}. Here ⟨options⟩ are the options passed to the to path. Inside ⟨path⟩ you can use the commands \tikztostart, \tikztotarget, and \tikztonodes. The value of \tikztostart is the start node and the value of \tikztotarget is the end node. The value of \tikztonodes is given by ⟨nodes⟩, i.e., the nodes of the to path. The values returned by \tikztostart and \tikztotarget do *not* include parentheses. ☑

Figure 5.33 defines a to path style called hvh that connects two points using three line segments. The first line segment is horizontal, the second is vertical, and the third is horizontal.

CHAPTER 6
Presenting Data in Tables

THIS CHAPTER STUDIES how to present data using tables. This includes studying the purpose of tables, table taxonomy, table anatomy, table design, wide and multi-page tables, and packages for spreadsheets and multi-page tables.

6.1 Why Use Tables?

Tables are a common way to communicate facts in newspapers, reports, journals, theses, and so on. Tables have several advantages.

- Tables list numbers in systematic fashion.
- Tables supplement, simplify, explain, and condense written material.
- Well-designed tables are easily understood.
    - ⋆ Patterns and exceptions can be made to stand out.
    - ⋆ They are more flexible than graphs. For example, in a graph it may be difficult to mix numeric information about data in different units such as the total consumption of petrol in Ireland in tons in the years 1986–2008 and the average number of rainy days per year in the same country.

6.2 Table Taxonomy

There are two kinds of tables: demonstration tables and reference tables. The following explains the difference between the two.

demonstration tables A demonstration table organises figures to show a trend or show a particular point. Examples are: (most) tables in technical reports, theses, and tables (shown) in meetings.

reference tables A reference table provides extra and comprehensive information. Examples are: train schedules, telephone directories, and stock market listings.

As a general rule a demonstration table in a thesis should be presented in the main text because you want to make a point with the information in the table. Likewise, a reference table in a thesis is probably best presented in an appendix because the data are "additional." However, exceptions are possible.

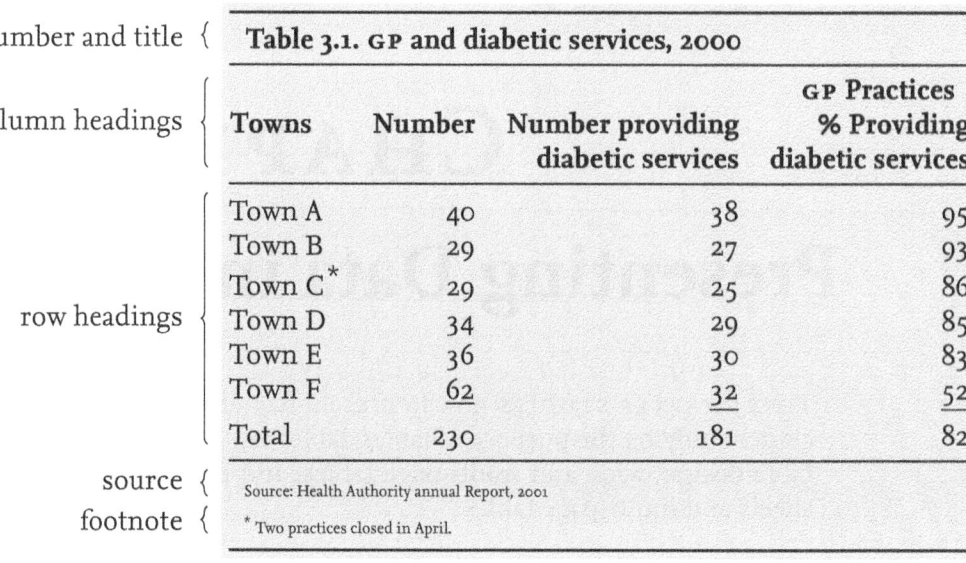

Figure 6.1
This figure shows the components of a typical demonstration table. The background of the table is coloured grey. The black-on-white text on the left of the table describes the components of the table.

number and title {

column headings {

row headings {

source {

footnote {

Table 3.1. GP and diabetic services, 2000			
Towns	**Number**	**Number providing diabetic services**	**GP Practices % Providing diabetic services**
Town A	40	38	95
Town B	29	27	93
Town C*	29	25	86
Town D	34	29	85
Town E	36	30	83
Town F	62	32	52
Total	230	181	82

Source: Health Authority annual Report, 2001

* Two practices closed in April.

6.3 Table Anatomy

Figure 6.1 depicts a typical demonstration table, which is based on [Bigwood, and Spore 2003, page 27]. The table has several components.

number and title In this example, the number and the title are listed at the top of the table. You may also find them at the bottom. The title should describe the purpose of the table. The table's number is used to reference the table further on in the text. It also helps locating the table.

There are also two other styles of tables. In the first you will find a separate *legend,* which is a description of what is in the table. In the other style, which is the default in LaTeX, tables have *captions,* which are a combination of number, title, and legend. Good captions should provide a number, a title, and a short explanation of the data listed in the table.

If you include a table, you should always discuss it in the text.
- If the table *is* relevant, does have a message, but is not referenced in the text, then how are you going to draw your reader's attention to the table? After all, you would want your reader to notice the table.
- If the table is *not* relevant to the running text, then why present it?
- If you don't discuss a table in the running text, then this may confuse and irritate the reader because they may waste a lot of time trying to find where this table is discussed in the text.

column headings The column headings are used to describe properties of the rows in the table. In this example, there are four column heading: 'Towns,' 'Number,' 'Number providing diabetic services,' and '% providing diabetic services.' Horizontal lines separate the column headings from the number and title and from the row heading of the table.

Chilled Meats	Calories
Beef (4 oz/100 g)	225
Chicken (4 oz/100 g)	153
Ham (4 oz/100 g)	109
Liver sausage (1 oz/25 g)	75.023
Salami (1 oz/25 g)	125

Table 6.1
A poorly designed table

row headings The row headings list what is in the rows of the table. In this table there are 7 row headings: 'Town A,' ..., 'Town F,' and 'Total.'

body Each row/column presents facts, patterns, trends, and exceptions in terms of numbers, and percentages.

trend In this table the general pattern is presented in the last column. Generally, in most towns more than 80% of the GPs provide diabetic services.

exception This table underlines exceptions of the general trend in the table. Other techniques for highlighting exceptions are using a different style of text (bold, italic, ...). However, notice that using different colours to highlight exceptions may not always be a good choice. For example, the difference may not be clear when the table is printed on a black-and-white printer. Also it may be difficult to reproduce colours with a photocopier.

source The source describes the base document from which the table is obtained or is based on.

footnote The footnote provides an additional comment about some of the data.

6.4 Table Design

This section provides some rules of thumb for the design of tables. To start, consider Table 6.1, which is based on [Bigwood, and Spore 2003, page 18]. It should be clear that this is a very poorly designed table. Several things are wrong with this table. The following are but a few.

- The gridlines make it difficult to scan the data in the table.
- The column alignment makes it difficult to compare the calories in the table.
- The figures (digits) have different widths. For example, the 1 is narrower than the 5. This makes it difficult to compare the numbers even if the numbers are properly aligned.
- The calories in the last column are for different weights of meat. For example, the 225 calories in the first row are for 100 g of meat but the 125 calories in the last row are for 25 g of meat. This makes it impossible to see the trend of calories per unit of weight. Of course the reader can work out the trend from the data in the first column, but that isn't their task. This is a common error: the information is in the table but it is poorly presented. As a result the table is useless.
- The precision of the data in the last column is different for different items. For example, for salami, it is listed with three decimals. It is

Table 6.2

An improved version of Table 6.1

Chilled Meats	**Calories per 100 g (4 oz)**
Salami	500
Liver sausage	300
Beef	225
Chicken	153
Ham	109

not clear why this is important and it only makes it more difficult to see the trend.

To improve the table we do the following.

○ We scale the numbers in the last column to the same unit: 100 g (4 oz). This has two advantages. First, we can leave out the weight information from the first column. Second, the numbers in the last columns are now in the same unit, which makes it easier to compare the numbers.
○ We reorder the rows to highlight the trend in the last column.
○ We reduce the grid lines to a minimum. This makes is easier to scan the data.
○ We present all numbers using the same precision and a similar number of digits. It is now easier to compare the numbers.
○ We align the items in the first column to the left. This now makes it easier scan the items in the first column.
○ We align the numbers to the right. This now makes it easier to see the relative differences of the data.
○ We use non-proportional (tabular) figures in the last column. Most LaTeX classes and styles already give you tabular figures. If they don't then they may have to enforce a command to switch to a style with tabular numbers. Alternatively, you can always get tabular figures with the command \texttt.
○ Optional: we may make the Column Headings stand out by typesetting them in a bold typeface.

The resulting table is listed as Table 6.2. Hopefully you agree this is a much better table.

The main rule of thumb in the design of tables is to keep them simple. Less is more.

○ Good tables are simple and uncluttered.
  ⋆ The number of vertical grid lines should be reduced to the absolute minimum. The advantage is that it makes it easier to scan the data in the table.
  ⋆ Other gridlines should be kept to a minimum. Arguably, gridlines should only be used to separate (1) the table from the surrounding text, (2) the number and title, (3) the column headings, and (4) the row headings.

○ Unless there is a good reason, you should align numbers to the right. Also it is usually better if you align all Column Headings to the right or align them all to the left. This results a more uniform appearance, which makes it easier to compare the numbers.

○ Good table titles should be concise, definitive, and comprehensive. Where appropriate they should inform the reader about the following.

what Table titles should describe the subject of table. For example: Annual income.

where If needed, they should describe the geographic location of the data.

when If needed, they should describe the relevant time. This should be kept short: 2000, 1900–1940, May, ...

units If units are used they should be described. Do not mix units, e.g., kilograms and pounds, because this makes it difficult to compare. Where possible, you should convert all numbers to the same unit (preferably, SI units).

○ Numbers should be aligned to *facilitate comparison*. For most reference tables and all demonstration tables:

⋆ Numbers should be typeset in a typeface with tabular (monospaced) figure (digit) glyphs. Alternatively, you may use the `dcolumn` or the `siunitx` packages. This is explained further on.

⋆ If a column consists of whole integer numbers then these numbers should be aligned to the right.

⋆ Decimal numbers should be aligned to the decimal point.

⋆ If there is much variance in the numbers, you could use scientific notation, e.g., `$1.4 10^{+4}$` and `$2.3 10^{-3}$`. Notice that the exponent may disrupt the normal inter-line spacing. Should this occur you may also consider using `1.4E␣+4` and `2.3E␣-3` or even `1.4E\,+4` and `2.3E\,-3`. If all signs of the exponent parts are nonnegative then consider omitting the signs.

⋆ If all numbers are of the same magnitude, consider scaling numbers to multiples of thousands, millions, ...

○ Reduce the amount of whitespace per line. This makes it easier to quickly scan the lines in the tables from left to right. With long lines and much whitespace this is much more difficult. In Table 6.2 the distance between the last letters in the first column and the first numbers in the second column is relatively small. Had we typeset the column heading of the second column on a single line (as opposed to using two lines) the distance would have been greater, leading to a less-quality table.

○ For long tables, you should consider adding extra linespace at regular intervals (for example after each fourth or fifth line).

6.5 Aligning Columns with Numbers

To produce useful tables, columns with numbers should be properly aligned. There are two approaches. The first approach is to use a typeface that has non-proportional (tabular) figures and align the numbers by hand. The second approach is to use a package such as the dcolumn and siunitx package.

6.5.1 *Aligning Columns by Hand*

If you're dealing with a few tables and if you haven't used packages to align your columns, then aligning the columns by hand is arguably the easier option. Since numbers should be aligned to right, the column alignments of the tabular environment should be r for all columns. If all your numbers in a column are integers or if they all have the same number of decimal digits, then life is easy. However, you must still make sure you use a font that has tabular figures. Most LaTeX fonts have tabular figures. Note that you can always get tabular figures if you use the command \texttt to typeset the numbers. In the following it is assumed the LaTeX font has tabular figures.

```
\begin{tabular}{rr}
  \toprule
    \textbf{Data}
  & \textbf{Data}
\\\midrule
    111 & 45.67
\\   45 & 56.78
\\\bottomrule
\end{tabular}
```

Data	Data
111	45.67
45	56.78

However, if the format of your data is less regular, then things get complicated. Still, you can always get the alignment perfect with the \hphantom command. (Remember that the command \hphantom{⟨stuff⟩} typesets ⟨stuff⟩ in invisible ink.) The following shows how to typeset the numbers in the table.

```
\begin{tabular}{rr}
  \toprule
    \textbf{Data}
  & \textbf{Data}
\\\midrule
    .2\hphantom{0} &  0.00
\\ 1.11            & 45.67
\\45.\hphantom{00} & 56.78
\\\bottomrule
\end{tabular}
```

Data	Data
.2	0.00
1.11	45.67
45.	56.78

Remember from Section 2.19.5 that @-expressions insert material between adjacent columns. In the following we exploit @-expressions to typeset the decimal dots. We may get our overall alignment correct if we align the digits to the left of the decimal point to the right and align the digits to the right of the decimal point to the left. Each number in the output table is now equivalent to two columns in the tabular environment in the input. Therefore, we need the \multicolumn command to typeset the column heading of the column.

```
\begin{tabular}{r@{.}lr@{.}l}
  \toprule
  \multicolumn
     {2}{r}{\textbf{Data}}
& \multicolumn
     {2}{r}{\textbf{Data}}
\\\midrule
   &2 & 0&00
\\ 1&11 & 45&67
\\45& & 56&78
\\\bottomrule
\end{tabular}
```

Data	Data
.2	0.00
1.11	45.67
45.	56.78

6.5.2 *The dcolumn Package*

The dcolumn package [Carlisle 2001] is intended for tabular or array environments. In the following we shall assume the package is used in a tabular environment. The purpose of the package is to provide support for numbers that are aligned on a decimal point symbol. The package provides new column alignment notation for the required argument of the tabular environment. Also it provides a command for defining shorthand notation for column alignment.

The package assumes the input uses an *input separator* that separates the whole numbers and the fractional part. If the target language is English then this will be a decimal point: 3.1415. Otherwise it may be a comma: 3,1415. It also assumes there is an *output separator* for the output. The following are the relevant commands.

D{⟨input separator⟩}{⟨output separator⟩}[⟨decimal places⟩]

Here ⟨input separator⟩ and ⟨output separator⟩ are the input and output separator symbols. Both should be a single character. The argument ⟨decimal places⟩ specifies the maximum number of decimal places in the input. If the value of this argument is negative then you may use any number of decimal places in the input. In addition, the numbers are centred about the ⟨output separator⟩ symbol. Providing negative values for ⟨input separator⟩ is discouraged because it may result in very wide columns. ☑

\newcolumntype{⟨shorthand⟩}{⟨column specifier⟩}

Here ⟨shorthand⟩ defines a new symbol for specifying column alignment in the required argument of the tabular environment. The argument ⟨column specifier⟩ specifies the column alignment, which is of the form D{⟨input separator⟩}{⟨output separator⟩}{⟨decimal places⟩}. ☑

\newcolumntype{⟨shorthand⟩}[1]{⟨column specifier⟩}

This is similar to the previous command but this time \newcolumntype takes an argument, which may be used in ⟨column specifier⟩. Using the notation #1 in ⟨column specifier⟩ gives you the value of the actual

Figure 6.2

Aligning columns with the dcolumn package

```
\newcolumntype%
    {p}[1]{D{.}{.}{#1}}
\begin{tabular}{p2p3p0}
  123.45 &  12.345 &  1 \\
  1234.5 & 123     & 12 \\
\end{tabular}
```

123.45	12.345	1
1234.5	123	12

Figure 6.3

Aligning columns with the siunitx package

```
\begin{tabular}{SS}
  123   &  23
\\  45. &    1.09
\\   .1 & 678.999
\\  7.7 &   1e10
\\ 33.3 &   2.2e-5
\end{tabular}
```

123	23
45.	1.09
.1	678.999
7.7	1×10^{10}
33.3	2.2×10^{-5}

argument. For example, with `\newcolumntype{p}[1]{D{.}{.}{#1}}` using p3 or p{3} is equivalent to {D{.}{.}{3}}. ☑

Figure 6.2 provides a short example of these commands.

6.5.3 *The siunitx Package*

Another package supporting number alignment in `tabular` environments is `siunitx` [Wright 2011]. It defines a new column alignment symbol S. By default this symbol aligns numbers to the decimal dot. Figure 6.3 presents an example. Note that the `siunitx` recognises "exponent notation" and expands it. More information about the package may be found in the package documentation [Wright 2011].

6.6 The table Environment

We have already seen how to present information in `tabular` environments. The `table` environment creates a *floating* table. As with the `figure` environment, this puts the body of the environment in a numbered table, which may be put in a different place in the document than where it's actually defined. The table placement is controlled with an optional argument. This optional argument works as with the optional argument of the `figure` environment. (See Section 4.1 for further information about the optional argument and how it affects the positioning of the resulting table.) Inside the table, `\caption` defines a caption, which also works as with `figure`. *Moving arguments* inside captions have to be protected. This is explained in Section 11.3.3. If you don't understand this, then don't worry, just write `\protect` before each command in the caption. Section 11.3.3 provides more details about moving arguments and protection. The starred version of the

```
\begin{table}[tbp]
   \begin{tabular}{ll}
   \toprule
       \textbf{Chilled Meats}
     & \textbf{Calories per} \\
     & \textbf{100\,g/4\,oz} \\
   \midrule
       ...
   \bottomrule
   \end{tabular}
   \caption[Calories of chilled meats.]
          {Calories of chilled meats per weight. ...
           \label{tab:meat}}
\end{table}
```

Figure 6.4
Creating a table

environment (table*) is for two-column documents. It produces a double-column table.

Figure 6.4 shows how to create a table. The example assumes the booktabs package is included.

6.7 Wide Tables

Sometimes tables are too wide for the current page. When this happens, use the rotating package. The package defines commands and environments that implement a sidewaystable environment, for presenting rotated tables. The following command typesets ⟨stuff⟩ in a rotated table.

```
\begin{sidewaystable}                                    LaTeX Usage
   ⟨stuff⟩
\end{sidewaystable}
```

Inside ⟨stuff⟩, the command \caption works as usual. The rotating package also defines a sidewaysfigure environment for figures.

6.8 Multi-page Tables

The longtable package defines an environment called longtable, which has a similar functionality as the tabular environment. The resulting structures can be broken by LaTeX's page breaking mechanism.

Since a single longtable may require several page breaks, it may take several runs before it is fully positioned. The \caption command works as usual inside the body of a longtable.

The longtable package needs to know how to typeset the first column heading, subsequent column headings, what to put at the bottom of the table on the last page, and what to put at the bottom of the first pages. This is done with the following commands.

\endfirsthead
This indicates the end of the first column headings specification. The

Figure 6.5

Using the longtable package. The \newcommand command at the top of the listing defines a user-defined command called \boldmc which works just as the \multicolumn command, except that it typesets the text in a bold face typeface. User-defined commands are explained in Chapter 11.

```
\newcommand\boldmc[3]{
    \multicolumn{#1}{#2}{\textbf{#3}}%
}
\begin{longtable}{lr}
    \toprule
    \textbf{Meats}
        & \boldmc{1}{l}{Calories per 100\,g}
    \\\midrule
\endfirsthead
    \toprule
    \boldmc{2}{c}{\tablename~\thetable\ Continued}
    \\\midrule
    \textbf{Meats}
        & \boldmc{1}{l}{Calories per 100\,g}
    \\\midrule
\endhead
    \midrule
    \boldmc{2}{l}{Continued on next page}
    \\\bottomrule
\endfoot
    \\\bottomrule
\endlastfoot
    Salami          & 500
    \\Liver sausage & 300
    ⋮
\end{longtable}
```

material from the start of longtable environment to the command \endfirsthead defines the first column headings.   ☑

\endhead

This indicates the end of the specification for the remaining column headings. All the material from \endfirsthead to \endhead is defines the the remaining column headings.   ☑

\endfoot

This indicates the end of the specification for remaining column headings. The material between \endhead and \endfoot is defines the bottom of the tables on pages all except for the last page.   ☑

\endlastfoot

This indicates the end of the last foot specification. The material between \endfoot and \endlatsfoot defines the bottom of the table on the last page.   ☑

Figure 6.5 typesets a long table with the longtable environment.

6.9 Databases and Spreadsheets

There are several packages that let you create and query databases and typeset the result as a table or tabular. Some of these packages provide additional functionality for creating barcharts, piecharts, and so on. The following are some of these packages.

datatool The datatool package [Talbot 2009] is a very comprehensive package. The package lets you create databases, query them, and modify them. The package also supports pie charts, bar charts, and line graphs. Further information may be found in the package documentation [Talbot 2009].

pgfplotstable The pgfplotstable package [Feuersänger 2010b] reads in tab-separated data and typesets the data as a tabular. The package also supports a limited form of queries. The package lets you round numbers up to a specified precision.

calctab The calctab package [Giacomelli 2009] lets you define rows in the table in terms of commands. The package can accumulate the sum of the entries in a given column, and so on. The package documentation is not very long and uses simple examples. Unfortunately the package does not seem to have a facility to set the symbol for the decimal point.

spreadtab As the name suggests, the spreadtab package [Tellechea 2011] is written with a spreadsheet in mind. The user specifies a matrix of cells (the spreadsheet) in some form of tabular-like environment. The layout of the matrix determines the result. Cells can contain input but they can also be rules for computing results from other cells. The package provides a command for setting the symbol for the decimal point.

If all you need is a simple way to compute sums/averages from rows and columns then you should consider using the spreadtab package.

CHAPTER 7
Presenting Data with Plots

THIS CHAPTER studies the presentation of data with "data plots" with LaTeX. Usually we shall use the word 'graph' instead of data plot. The presentation of this chapter is example driven and mixes theory of presentation with practice. The theory is mostly based on [Bigwood, and Spore 2003]. With the exception of pie charts, all graphs were created with the pgfplots package [Feuersänger 2010a], which creates astonishingly beautiful data plots in a consistent style with great ease. The remainder of this chapter covers pie charts, bar graphs, paired bar graphs, component bar graphs, line graphs, and scatter plots but it excludes 3-dimensional data plots.

7.1 The Purpose of Data Plots

A picture can say more than a thousand words. This, to some extent, epitomises data plots and graphs: good data plots tell a story that is easily recognised and will stick. Plots are good at showing global trends and relationships, differences, and change.

global trends Graphs are good at showing global trends and relationships. A 2-dimensional scatter plot, for example, may reveal that the data are clustered, that the y-coordinates have a tendency to increase as the x-coordinates increase, that most x-coordinates are smaller than the y-coordinates, and so on.

differences Graphs are good at showing differences between two or several functions/trends. For example, the difference between the height of males and females as a function of their age, the difference of the running times of four algorithms as a function of the input size, and so on.

rate of change Graphs are also good at showing the rate of change within a single function/trend. For example, the rate of change of the running time of a single algorithm as the size of the input increases. Interestingly, differences and change can often be shown effectively in a single graph.

Well-designed graphs stick and convey the essence of the relationship.

7.2 Pie Charts

Our first kind of graph is the pie chart. Pie charts have become very fashionable. Programs such as Excel have made creating pie charts

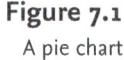
Figure 7.1
A pie chart

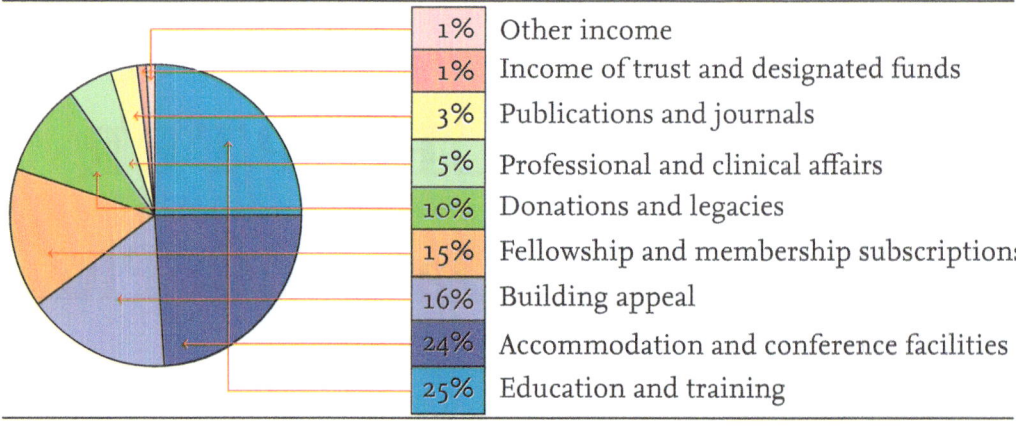

1%	Other income
1%	Income of trust and designated funds
3%	Publications and journals
5%	Professional and clinical affairs
10%	Donations and legacies
15%	Fellowship and membership subscriptions
16%	Building appeal
24%	Accommodation and conference facilities
25%	Education and training

relatively easy. Figure 7.1 depicts a typical pie chart. The information in the pie chart is adapted from [Bigwood, and Spore 2003]. For sake of this example, the percentages are listed as part of the legend information. Even with this information it is difficult to relate the segments in the chart with the items in the legend.

The relative size of each segment in the pie chart is equal to the relevant size of the contribution of its "label." To create the chart, the segments are ordered from small to large and presented counter-clockwise, starting at 90°. Colours are usually used to distinguish the segments. Note that care should be taken when selecting colours because they may not always print well.[1] Hatch patterns should be avoided because they distract. The pie chart in Figure 7.1 has 9 segments, which is too much: good pie charts should have no more than 5 segments [Bigwood, and Spore 2003].

Note that without the percentages it is impossible to compare the relative sizes/contributions of the two smaller segments, which happen to have the same size. Likewise it is difficult to compare the sizes of the segments that contribute 15% and 16% of the total. Arguably, a table is a better way to present the data. As a matter of fact, the legend is already a table and the combination of the legend and pie chart is redundant.

Pie charts are very popular, especially in "slick" presentations. This is surprising because it is well known that pie charts are not very suitable for communicating data and that specialists avoid them [Bigwood, and Spore 2003]. Tantau [2010] also presents some good arguments against pie charts because they may distort the information. Bar graphs, which are studied in the following section, are almost always more effective than pie charts. Despite these observations, pie charts are good at showing parts of a whole by percentages, and how a few components may make up a whole [Bigwood, and Spore 2003].

---

[1]Note that with careful planning you should be able to change the colours of the segments depending on a global "mode" settings in your document. Specifically, this should allow you to select different, proper colours for an online version and a printable version of your document. Techniques for changing colours and other setting depending on modes are studied further on in this book.

```
\begin{tikzpicture}
\begin{axis}[width=8cm, height=6cm, tick align=outside]
   \addplot[draw=blue]
          coordinates {(0,1) (1,1) (2,3) (3,2) (4,2)};
   \addlegendentry{Line 1}
   \addplot[draw=red]
          coordinates {(0,0) (1,4) (2,4) (3,3) (4,3)};
   \addlegendentry{Line 2}
\end{axis}
\end{tikzpicture}
```

Figure 7.2

Using the axis environment

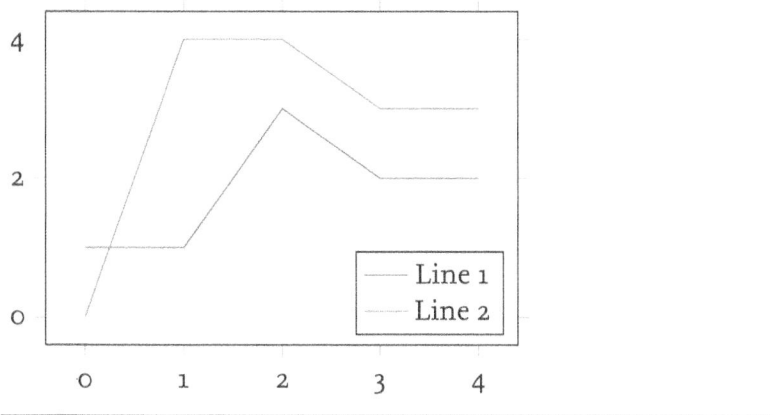

Figure 7.3

Output of the axis environment in Figure 7.2.

The pie chart in Figure 7.1 was drawn using raw tikz commands and drawing the chart took a long time. Nicola Talbot's csvpie package provides some support for drawing pie charts but be aware of the arguments against pie charts.

In the remainder of this chapter we shall omit the tikzpicture environment from all examples.

7.3 Introduction to pgfplots

This section provides an introduction to drawing graphs with the pgfplots package that is built on top of pgf. The pgfplots package lets you draw graphs in a variety of formats. The resulting graphs have a consistent, professional look and feel. The package also lets you import data from matlab. As is usual with the pgf family, their manuals are impressive [Feuersänger 2010a].

The workhorse of the pgfplots package is an environment called axis, which may define one or several *plots* (graphs). Each plot is drawn with the command \addplot. When the graphs are drawn the environment also draws a 2- or 3-dimensional axis. The axis environment is used inside a tikzpicture environment, so you can also use tikz commands. The options of the axis environment specify the type of the plot, the width, the height, and so on. Figure 7.2 demonstrates how to use the command and Figure 7.3 depicts the resulting output. The example is not intended to be pretty.

As is hopefully clear from this example, the options of the `axis` environment set the width to 8 cm, set the height to 6 cm, and force the ticks to be on the outside of the axes.

If you always present your `axis` environments with the same settings then your graphs will look consistent. For example, you probably want to use a default height and width for your graphs. To save work you'd like to define default values for height and width and omit the height and width specifications in the `axis` environment, except when overriding them. This is where the command `\pgfplotsset` comes into play. Basically, `\pgfplotsset` is to `pgfplots` what `\tikzset` is to `tikz`: it lets you set the default values for options of `pgfplots` commands and environments. The following is an example.

```
\pgfplotsset{width=6cm,height=4cm,
             compat=newest,
             enlargelimits=0.18}
```
LATEX Input

This example sets the default compatibility to `newest`, which is advised. The default width is set to 6 cm and the default height is set to 4 cm. The spell `enlargelimits=0.18` increases the default size of the axes by 18 %. As with `tikz` commands, you may override the values for these options by passing them to the optional arguments of the `axis` environment and the `\addplot` command.

7.4 Bar Graphs

Our next graph is the bar graph. Bar graphs present quantities as rows or columns. You can use bar graphs to show differences, rates of change, and parts of a whole.

Figure 7.4 depicts a typical bar graph. As with rows in a table, the bars of the graph are ordered to show the main trend. Notice that the data in this graph could just as well have been presented with a table. However, the main advantage of the bar graph presentation is that it "sticks" better. For example, it is very clear from the shape of the bars that Kilkenny, Cork, and Tipperary are the main all-Ireland hurling champions. It is also clear these teams are much better than the rest. With a table, the impact of the difference would not have been so big. Also notice that even in the absence of the frequency information after the bars, it is relatively easy to compare relative differences between the bars.

It is also possible to have bar graphs with vertical bars. Such bar graphs are sometimes used to present changes over time. For example, if you use *x*-coordinates for the time, use *y*-coordinates for the quantities, and order the bars by time from left to right, then you can see the rate of change of the quantities over time. Of course, you can also present changes over time with horizontal bar graphs but some people find this intuitively less pleasing.

There are at least two reasons why vertical bar graphs are not as

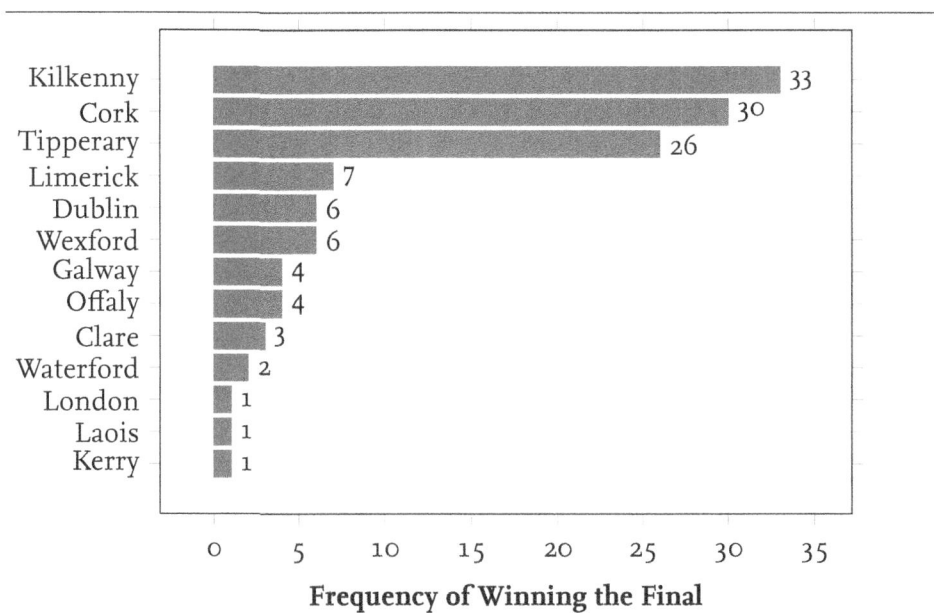

Figure 7.4

A bar graph. All-Ireland hurling champions and the number of times they've won the title before 2012.

```
\begin{axis}
    [xbar,width=11cm,height=8cm,bar width=10pt,enlargelimits=0.13,
     nodes near coords,
     nodes near coords align=horizontal,
     point meta=x * 1, % The displayed number.
     legend pos=south east,
     xlabel=\textbf{Frequency of Winning the Final},
     tick align=outside,
     xtick={0,5,...,35}, ytick={1,...,13},
     yticklabels={Kerry,Laois,London,Waterford,Clare,Offaly,Galway
                  Wexford,Dublin,Limerick,Tipperary,Cork,Kilkenny}]
\addplot[draw=blue, fill=blue!15] coordinates
    {(1,1) (1,2) (1,3) (2,4) (3,5) (4,6) (4,7)
     (6,8) (6,9) (7,10) (26,11) (30,12) (33,13)};
\end{axis}
```

Figure 7.5

Creating a bar graph

ideal as horizontal bar graphs. First it makes it difficult to label the bars, especially if the text of the labels is long. For example, putting the labels along the *x*-axis usually requires rotating the labels, which makes it difficult to read the labels. Second, you can put more bars in a graph with horizontal bars.

Figure 7.5 presents the input for the bar graph in Figure 7.4. The graph itself is typeset inside an axis environment, which itself is inside a tikzpicture environment—remember the tikzpicture environment is omitted. The xbar option of the axis environment specifies that this is a horizontal bar graph. The data for the graph are provided by the \addplot command. The enlargelimits=0.13 option is used to increase the size of the axes by 13 %.

The `xtick` and `ytick` keys override the default positions of the x and y ticks on the axes. For each `xtick` (`ytick`) position, `pgfplots` will place a little tick at the position on the x-axis (y-axis) and label the tick with its position. The tick labels can be overridden by providing an explicit list. This is done with the commands `\xticklabels` and `\yticklabels`. The input in Figure 7.5 uses the command `\yticklabels` to override the labels for the y ticks. The left-to-right order of the labels in the argument of `\yticklabels` is the same as the increasing order of the y ticks in the bar plot. The command `\xticklabels` works in a similar way.

The lengths of the bars are defined by the required argument of the `\addplot` command. The length of the bar with y-coordinate y is set to x by adding the tuple (x, y) to the list. For example, the tuple `(3r,13)` defines the length of the bar for Kilkenny.

The bar graph in Figure 7.4 also lists the lengths of the bars. These lengths are typeset with the keys `nodes near coords`, `nodes near coords align`, and `point meta`. By default the lengths of the bars are not printed. The `point meta` key defines the values near the bars. Since we want to typeset the length of the bar, which is defined by the x-coordinate in the coordinate list, we use `x * 1`. More complex expressions are also allowed.

7.5 Paired Bar Graphs

Paired bar graphs are like bar graphs but they present information about two groups of data. Figure 7.6 depicts an example of a paired bar graph. In this graph there are two bars for each of the five experiments: one for FC and one for MAC. The information in the graph is made up.

Before studying the LaTeX input, it should be noticed that `pgfplots` also lets you construct similar graphs with more than two groups of data. Also notice that such graphs should be discouraged because the number of bars soon becomes prohibitive, making it difficult to see the trends.

Figure 7.7 shows the input for the horizontal paired bar graph from Figure 7.6. As you can see from the input, a horizontal paired graph is also created by passing `xbar` as an option to the `axis` environment. The rest of the input is also similar to the input we needed to define our horizontal bar graph. The main differences are that (1) we have two bar classes per y-coordinate and (2) we have a legend. For each bar class there is an entry in the legend.

Each class of bars is defined by a separate call to `\addplot`. The command `\addlegendentry` adds an entry to the legend for the most recently defined class. The style of the legend entries is set with the `area legend` option, which results in a rectangle drawn in the same way as the corresponding bar. This is slightly nicer than the default legend entry style.

The style of the legend is set with the `legendstyle` key. The `legend`

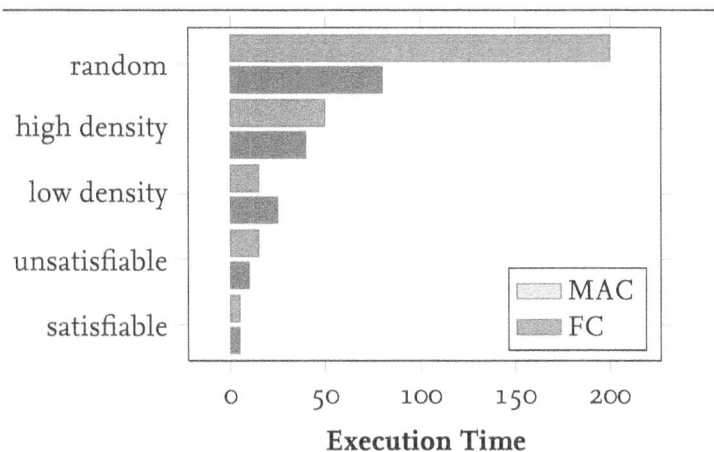

Figure 7.6

A paired bar graph. Comparison of the execution time of the Maintain Arc Consistency (MAC) and the Forward Checking (FC) algorithms for instances of 5 problem classes. Execution time in seconds.

```
\begin{axis}
    [xbar,enlargelimits=0.14,width=8cm,height=6cm,,
    bar width=10pt,area legend,legend pos=south east,
    legend style={legend pos=north east,
                 cells={anchor=west}},
    tick align=outside,xlabel=\textbf{Execution Time},
    ytick={1,...,5},
    yticklabels={satisfiable,unsatisfiable,
                 low density,high density,random}]
\addplot[draw=blue,fill=blue!15]
    coordinates {(5,1) (10,2) (25,3) (40,4) (80,5)};
\addlegendentry{\textsc{MAC}}
\addplot[draw=blue,fill=blue!50]
    coordinates {(5,1) (15,2) (15,3) (50,4) (200,5)};
\addlegendentry{\textsc{FC}}
\end{axis}
```

Figure 7.7

Creating a paired bar graph

pos key is used to position the legend. The spell cells={anchor=west} aligns the labels of the legend on the left.

7.6 Component Bar Graphs

Component bar graphs, also known as stacked bar graphs, compare several classes of data. Each class consists of (the same) components and within each class you can see the contribution of the components to the class as a whole. Figure 7.8 depicts a component bar graph.

Again notice that the bars are ordered to show the trend. For medal rankings, the first criterion is the number of gold medals won. Ties are broken by considering the number of silver medals, and so on. For other data you may have to order your rows depending on the overall size of the bars.

For the medal ranking example, it is easy to compare the contribution of the different medals to the overall medal count of a given country. Likewise, it is easy to compare the number of gold, silver, or bronze medals won by different countries. The reason why this works is that all sizes are small and discreet. For different kinds of data, with large ranges of data values, comparing the component sizes is usually not so easy.

Bigwood, and Spore [2003] discourage component bar graphs because they easily distort data and usually contain too much information.

Tables may be an interesting and good alternative to component bar graphs. For example, you can have a different row heading for each component in the component graph. If the total size of the bars is important then you can introduce a separate row heading to present these data as the "grand total," or as the "total time," and so on.

Figure 7.9 is the input for the component bar graph depicted in Figure 7.8. The options `xbar stacked` and `stack plots=x` indicate that the plot is a horizontal component bar graph. Each `\addplot` command defines the contribution of the next horizontal component for each y-tick position, so `(1,2)` in the argument of the first `\addplot` command states that The Netherlands (2) won one (1) gold medal. Likewise `(0,3)` in the argument of the second `\addplot` command states that France won no silver medals.

7.7 Coordinate Systems

None of our previous `pgfplots`-drawn graphs required additional `tikz` commands for additional lines or text. However, graphs with additional text and lines are quite common. The `pgfplots` package provides several dedicate coordinate systems for correctly positioning such additional text and lines. The following are some of these coordinate systems.

axis cs This system coincides with the numbers on the axes. Each coordinate in this system has the same x and y coordinates as the coordinates in the `\addplot` command. For example, if you use the command `\addplot{(1,2) (3,4)}` then the command `\tikz \draw (axis cs:1,2) node {⟨text⟩};` should draw ⟨text⟩ at the first coordinate.

rel axis cs This coordinate system uses coordinates from the unit square and linearly transforms them to plot coordinates. In this coordinate system the coordinates $(0,0)$ and $(1,1)$ are the lower left and the upper right corners of the unit square, so `\tikz \draw (rel axis cs:0.5,0.5) node {⟨text⟩};` should draw ⟨text⟩ in the centre of the plot.

xticklabel cs This coordinate system is for coordinates along the x-axis. Basically, the coordinate `xticklabel cs:x` is equivalent to `rel axis cs:x,0`. So far, this is not very interesting. However, the coordinate system also lets you provide an additional coordinate, which should be a length. When provided, the length defines the distance of a shift "away" from the labels on the x-axis.

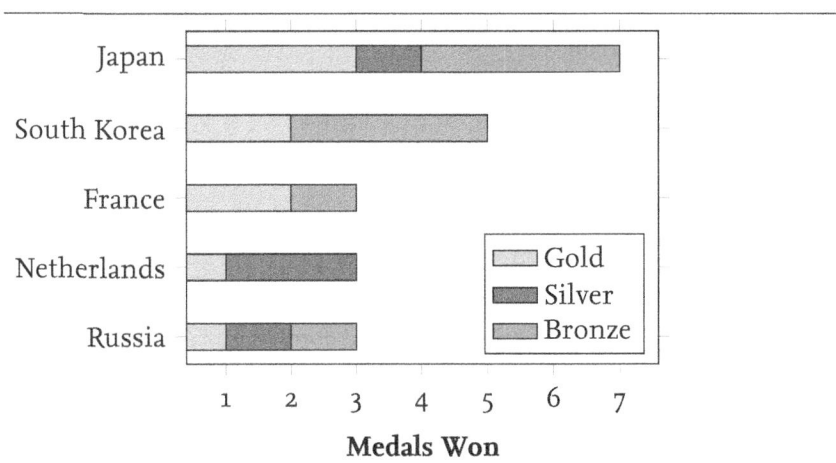

Figure 7.8
A component bar graph. Top five countries of the medal ranking of the 2009 World Judo Championships in Rotterdam (The Netherlands). (Source wikipedia.)

```
\begin{axis}
    [xbar stacked, stack plots=x, tick align=outside,
    width=8cm, height=6cm, bar width=10pt,
    legend style={cells={anchor=west}}, area legend,
    xlabel=\textbf{Medals Won}, ytick={1,...,5},
    yticklabels={Russia,Netherlands,France,
                 South Korea,Japan}]
\addplot[draw=black,yellow!50!brown]
    coordinates {(1,1) (1,2) (2,3) (2,4) (3,5)};
\addlegendentry{Gold}
\addplot[draw=black,white!60!gray]
    coordinates {(1,1) (2,2) (0,3) (0,4) (1,5)};
\addlegendentry{Silver}
\addplot[draw=black,orange!70!gray]
    coordinates {(1,1) (0,2) (1,3) (3,4) (3,5)};
\addlegendentry{Bronze}
\end{axis}
```

Figure 7.9
Creating a component bar graph

yticklabel cs This coordinate system is for coordinates along the *y*-axis. It works similar to the xticklabel cs coordinate system.

The remaining sections provide examples that use some of these coordinate systems. The reader is referred to the pgfplots package documentation [Feuersänger 2010a] for further information.

7.8 Line Graphs

Line graphs are ideal for presenting differences between data sets and presenting the rate of change within individual data sets. They are commonly used to present data (observations) that are a function of (depend on) a given parameter. For example, the running time of a

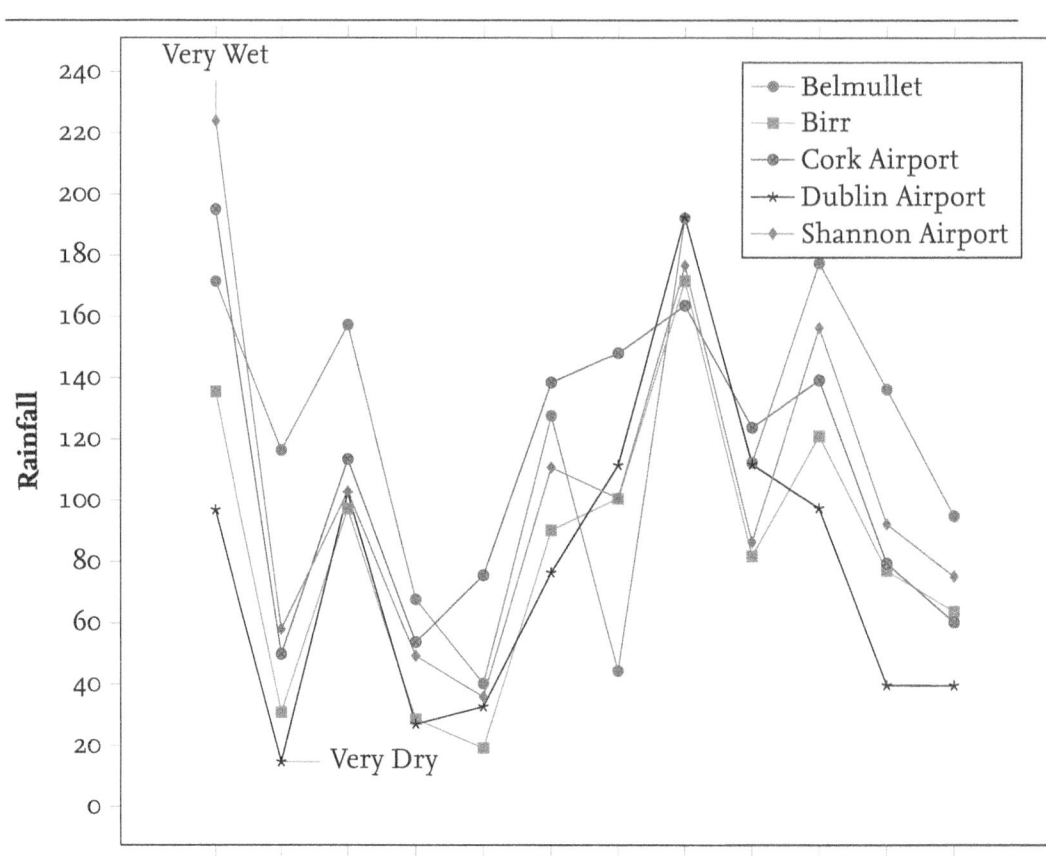

given algorithm as a function of the input size, the average height of males as a function of their age, and so on.

Figure 7.10 depicts a typical line graph. The legend in the top right hand corner of the graph labels the line types in the graph. In general legends should be avoided: if possible the lines should be directly labelled [Tufte 2001; Bigwood, and Spore 2003], which is to say that each label should be near its line. The main motivation for direct labelling is that legends distract and make it more difficult to relate the lines and their labels. Direct labelling usually makes it easier to relate the lines and their labels/purpose. For the graph in Figure 7.10 direct labelling is virtually impossible.

The x-tick labels in Figure 7.10 are positioned poorly. For example, the label Jan appears higher than the label Feb. This is probably caused by the fact that Feb doesn't have descenders. This is easily fixed by adding a \vphantom{g} command or a \strut command to the labels.

Figure 7.11 depicts the input for the line graph in Figure 7.10. Most of this is pretty straightforward. The command \addplot+ is used to define the lines in the graph. The extra plus in the command results in extra marks on the lines for the coordinates in the required argument of \addplot+. The option sharp plot of the \addplot command states that consecutive points in the plot should be connected using a straight

```
\begin{axis}
    [width=\textwidth,enlargelimits=0.13,tick align=outside,
     legend style={cells={anchor=west},legend pos=north east},
     xticklabels={Jan,Feb,Mar,Apr,May,Jun,Jul,Aug,Sep,Oct,Nov,Dec},
     xtick={1,2,3,4,5,6,7,8,9,10,11,12},
     xlabel=\textbf{Month}, ylabel=\textbf{Rainfall}]
\node[coordinate,pin=above:{Very Wet}] at (axis cs:1,223.9) {};
\node[coordinate,pin=right:{Very Dry}] at (axis cs:2,14.7)  {};
\addplot+[sharp plot] coordinates
        {(1,171.5) (2,116.4) (3,157.4) (4,67.7) (5,40.2) (6,127.6)
         (7,44.3) (8,192.1) (9,112.4) (10,177.5) (11,136.2) (12,94.8)};
\addlegendentry{Belmullet}
\addplot+[sharp plot] coordinates
        {(1,135.5) (2,30.8) (3,97.3) (4,28.6) (5,19.2) (6,90.2
         (7,100.6) (8,171.6) (9,81.8) (10,121.0) (11,77.0) (12,63.7)};
\addlegendentry{Birr}
\addplot+[sharp plot] coordinates
        {(1,195.1) (2,49.8) (3,113.5) (4,53.7) (5,75.6) (6,138.5
         (7,148.1) (8,163.6) (9,123.8) (10,139.2) (11,79.4) (12,60.2)};
\addlegendentry{Cork Airport}
\addplot+[sharp plot] coordinates
        {(1,96.9) (2,14.7) (3,102.4) (4,27.0) (5,32.7) (6,76.4
         (7,111.5) (8,192.4) (9,111.8) (10,97.4) (11,39.6) (12,39.5)};
\addlegendentry{Dublin Airport}
\addplot+[sharp plot] coordinates
        {(1,223.9) (2,58.0) (3,102.9) (4,49.2) (5,35.9) (6,110.8
         (7,100.8) (8,176.6) (9,86.4) (10,156.4) (11,92.2) (12,75.1)};
\addlegendentry{Shannon Airport}
\end{axis}
```

Figure 7.11

Creating a line graph

line segment. This is more than likely what you want when you're creating line graphs. The \node commands at the end of the axis environment draw the texts 'Very Wet' and 'Very Dry' using the axis cs coordinate system. The node shape pin is new but it should be clear how it works.

Finally, notice that line graphs have a tendency to become crowded as the number of lines increases. If this happens, you should consider reducing the number of lines in your graph. Alternatively, you may consider using the spy feature to zoom in on the important crowded areas in you graph. The spy mechanism is explained in Section 5.10.

7.9 Scatter Plots

Scatter plots are ideal for discovering relationships among a huge/large set of 2-dimensional data points. Basically, the plot has a mark at each coordinate for each data point. Figure 7.12 is a scatter plot that compares the running times of two algorithms for different inputs. For each input, i, the scatter plot has a point at position (x_i, y_i), where

Figure 7.12

A scatter plot. Running time of Algorithm 1 versus running time of Algorithm 2. Running times in seconds. The majority of the coordinates are below the line $x = y$. This shows that Algorithm 1 requires more time for most input.

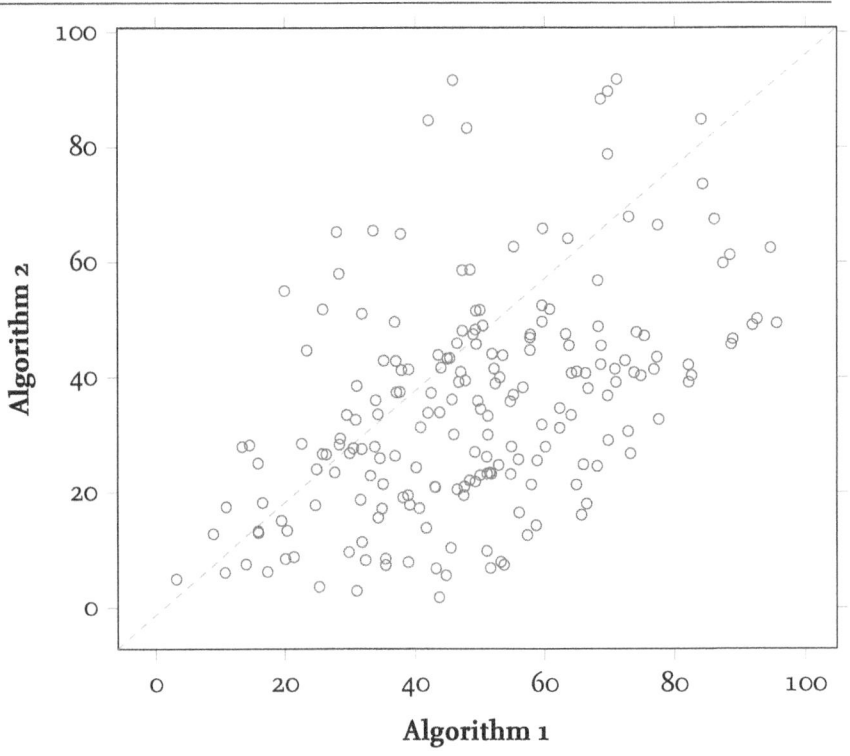

Figure 7.13

Creating a scatter plot

```
\begin{axis}
      [width=\textwidth, tick align=outside,
       xlabel=\textbf{Algorithm~1},
       ylabel=\textbf{Algorithm~2}]
\addplot{scatter,only marks,mark=o,
        draw=blue,scatter src=explicit}
        file {data.dat};
\draw[dashed,red!40]
      (rel axis cs:0,0) -- (rel axis cs:1,1);
\end{axis}
```

x_i is the running time of the first algorithm for input i and y_i is the running time of the second algorithm for input i.

As you can see from the scatter plot, Algorithm 1 usually takes more time than Algorithm 2 for random input because most points in the scatter plot have larger x-coordinates than y-coordinates: most points are below the line $x = y$. Furthermore, the overall shape of the plot suggests that the running times are positively correlated. The dashed red line helps detecting both trends in the plot.

Figure 7.13 presents the code for the scatter plot in Figure 7.12. The option scatter states that the coordinates provided by the calls to \addplot are for a scatter plot. The option only marks results in a mark at each coordinate that is specified by \addplot. The style of the mark may be set with the style mark=⟨mark style⟩. Possible values for

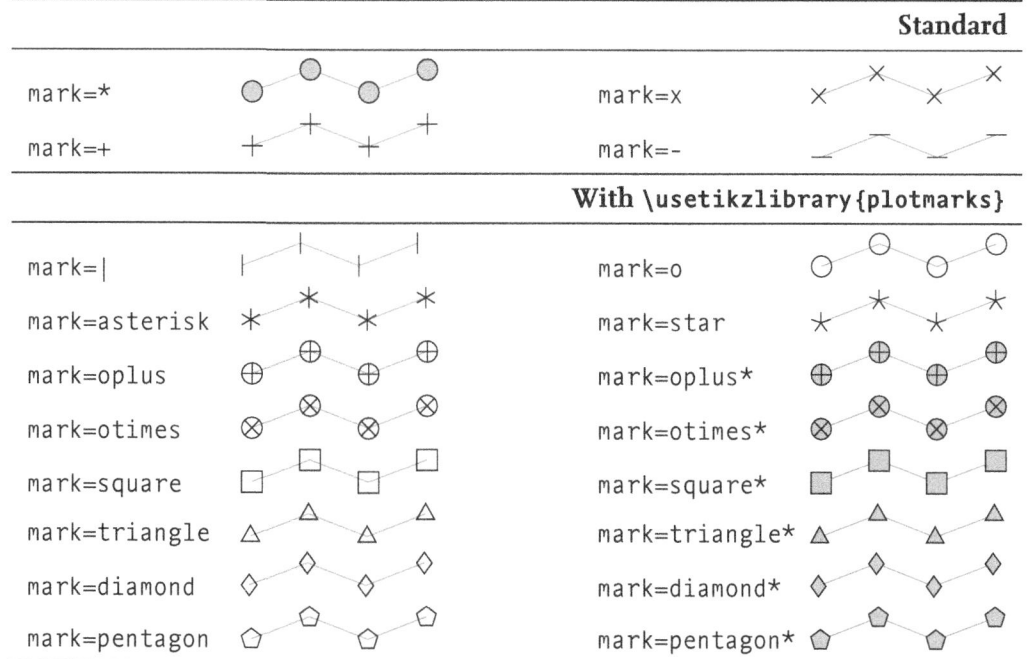

		Standard		
mark=*	⊙ ⊙ ⊙ ⊙		mark=x	✕ ✕ ✕ ✕
mark=+	+ + + +		mark=-	— — — —
		With \usetikzlibrary{plotmarks}		
mark=|	⊢ ⊢ ⊢ ⊢		mark=o	○ ○ ○ ○
mark=asterisk	✳ ✳ ✳ ✳		mark=star	✦ ✦ ✦ ✦
mark=oplus	⊕ ⊕ ⊕ ⊕		mark=oplus*	⊕ ⊕ ⊕ ⊕
mark=otimes	⊗ ⊗ ⊗ ⊗		mark=otimes*	⊗ ⊗ ⊗ ⊗
mark=square	□ □ □ □		mark=square*	▪ ▪ ▪ ▪
mark=triangle	△ △ △ △		mark=triangle*	▲ ▲ ▲ ▲
mark=diamond	◇ ◇ ◇ ◇		mark=diamond*	◆ ◆ ◆ ◆
mark=pentagon	⬠ ⬠ ⬠ ⬠		mark=pentagon*	⬟ ⬟ ⬟ ⬟

Table 7.1
This table lists the values for the mark option. The options at the top of the table are standard. The remaining options rely on the tikz library plotmarks.

⟨mark style⟩ and the resulting marks are listed in Table 7.1. In our example we're using the style o, which results in a circle. The option color=⟨colour⟩ sets the colour of the mark.

The option scatter src=explicit symbolic states that the co-ordinates are expected as explicit coordinates. Usually scatter plots have many data points. Adding all point specifications to the main LaTeX source of your pgfplot environments surely doesn't make it easier to maintain the environments. This is why pgfplots provides support for including data from external source files. In our example, file {data.dat} indicates that the coordinates are in the external file data.dat. All lines in this file are of the form ⟨x-coordinate⟩␣⟨y-coordinate⟩.

The red dashed line is drawn at the end of the axis environment. The rel axis cs coordinate system is used to specify the start and endpoint of the line. It is recalled from Section 7.7 that this coordinate system scales all coordinates to the unit square with lower left coordinate (0, 0) and upper right coordinate (1, 1).

PART IV

Mathematics and Algorithms

Oil paint and charcoal on canvas (05/09/09), 152 × 213 cm
Work included courtesy of Billy Foley
© Billy Foley (www.billyfoley.com)

CHAPTER 8
Mathematics

THIS CHAPTER is an introduction to typesetting basic mathematics in LaTeX. The following chapter is an introduction to typesetting more advanced mathematics. Further information about mathematical typesetting in LaTeX may be found in a book such as [Lamport 1994], a tutorial such as [Oetiker et al. 2007], or a book on using LaTeX for writing mathematics [Voß 2010]. A comprehensive listing of LaTeX symbols, including math symbols, is provided by Pakin [2005].

LaTeX's basic support for mathematics is limited, which is why the American Mathematical Society (AMS) provides a package called amsmath, which redefines some existing commands and environments and provides additional commands and environments for mathematical typesetting. Throughout this chapter it is assumed that you have installed the amsmath package.

8.1 The $\mathcal{AMS}$-LaTeX Platform

$\mathcal{AMS}$-LaTeX is a useful platform for typesetting mathematics. The software is provided by the AMS (http://ams.org/). The software is freely available and should come with any good LaTeX distribution. You can download the AMS software and documentation from http://www.ams.org/tex/amslatex.html.

The software distributed under the name $\mathcal{AMS}$-LaTeX consists of various extensions for LaTeX. The distribution is divided into two parts:

amscls The amscls class provides the AMS document class and theorem package. Using this class gives your LaTeX document the general structure and appearance of an AMS article or book.

amsmath The amsmath package is an extension package that provides facilities for writing math formulas and improving the typography.

Throughout this chapter $\mathcal{AMS}$-LaTeX and amsmath are used interchangeably. The amsmath package is really a collection of packages. If you include amsmath then you include them all. The amsmath package also provides support for configuring basic document settings. As usual this is done by passing options inside the square brackets of the \usepackage command: \usepackage[⟨options⟩]{amsmath}. Some of the options are as follows.

leqno Place equation numbers on the left.

reqno Place equation numbers on the right.

fleqn Position equations at a fixed indent from the left margin.

Some of the packages provided by $\mathcal{AMS}$-LaTeX are listed next. The description of the packages is adapted from the $\mathcal{AMS}$-LaTeX documentation [AMS 2002].

amsmath Defines extra environments for multi-line displayed equations, as well as a number of other enhancements for math (including the amstext, amsbsy, and amsopn packages).

amstext Provides a \text command for typesetting text inside a formula.

amsopn Provides the \DeclareMathOperator command for defining commands to typeset functions and operators such as \sin and \lim.

amsthm Provides a proof environment and extensions for the \newtheorem command.

amscd Provides an environment for simple commutative diagrams.

amsfonts Provides extra fonts and symbols, including boldface (\mathbf), blackboard boldface (\mathbb), and fraktur (\mathfrak).

amssymb Provides lots of extra symbols.

8.2 LaTeX's Math Modes

LaTeX has three basic modes that determine how it typesets its input. These modes are:

text mode In this mode the output does not have mathematical content and is typeset as text. Typesetting in text mode is explained in Chapter 1.

ordinary math mode In this mode the output has mathematical content and is typeset in the running text. Ordinary math mode is more-commonly referred to as inline math mode.

display math mode In this mode the output has mathematical content and is typeset in a display.

The mechanism for typesetting mathematics in ordinary (inline) math mode is explained in the following section. This is followed by some sections explaining some basic math mode typesetting commands, which are then used in Section 8.6. The main purpose of Section 8.6 is to describe some environments for typesetting display math.

8.3 Ordinary Math Mode

This section explains how to typeset mathematics in ordinary (inline) math mode. It is recalled from the previous section that this means that the resulting math is typeset in the running text. Typesetting in display math mode is postponed until Section 8.6. The $ operator switches from text mode to ordinary math mode and back, so $a = b$ results in $a = b$ in the running text. The following provides another example. If you don't understand the constructs inside the $\cdot$ $ expressions then don't worry: they are explained further on.

```
The Binomial Theorem states
$\sum^{n}_{i=0}
  \binom{n}{i} a^{i} b^{n-i}
  = (a + b)^{n}$.
Substituting $1$ for $a$
  and $1$ for $b$ gives us
$\sum^{n}_{i=0}
  \binom{n}{i} = 2^{n}$.
```

The Binomial Theorem states $\sum_{i=0}^{n} \binom{n}{i} a^i b^{n-i} = (a+b)^n$. Substituting 1 for a and 1 for b gives us $\sum_{i=0}^{n} \binom{n}{i} = 2^n$.

The mathematical expressions in the output are typeset in the running text. This should not come as a surprise because $\cdot$ $ is for typesetting in ordinary (inline) math mode.

8.4 Subscripts and Superscripts

Subscripts and superscripts are ubiquitous in mathematics. We've already seen subscripts and superscripts in some of the examples. This section formally explains how to use them.

The superscript operator (ˆ) creates a superscript. The expression $\langle expr \rangle^\langle sup \rangle$ makes $\langle sup \rangle$ a superscript of $\langle expr \rangle$. So `$x^2 + 2 x + 1$` gives you $x^2 + 2x + 1$. Grouping works as usual. So to typeset e^{a+b} you need braces: `$e^{a+b}$`.

Subscripts are handled as superscripts. The subscript operator (_) creates a subscript. The expression $\langle expr \rangle_\langle sub \rangle$ makes $\langle sub \rangle$ a subscript of $\langle expr \rangle$. So to get $f_{n+2} = f_{n+1} + f_n$ you need `$f_{n + 2} = f_{n + 1} + f_n$`.

Subscripts and superscripts may be nested and combined. The expressions $\langle expr \rangle_\langle sub \rangle^\langle sup \rangle$ and $\langle expr \rangle^\langle sup \rangle_\langle sub \rangle$ are equivalent and make $\langle sub \rangle$ a subscript of $\langle expr \rangle$ and $\langle sup \rangle$ a superscript of $\langle expr \rangle$, so `$s^{m + 1}_{n+2}$` gives you s_{n+2}^{m+1}.

It is good practice to avoid subscripts and superscripts inside subscripts and superscripts—some style and class files may reject them. The following are some advantages.

simplicity The fewer the subscripts and superscripts, the simpler the notation, the greater the transparency.
readability The resulting expression is easier to parse.
spacing Nested subscripts and superscripts may result in inconsistencies in the interline spacing, which doesn't look good. Avoiding nested subscripts and superscripts avoids such inconsistencies.

8.5 Greek Letters

This section describes the commands for typesetting Greek letters in math mode. These commands do *not* work in text mode.

Table 8.1

This table lists the math mode commands for lowercase Greek letters. The commands at the top of the table are standard LATEX commands. The command \digamma and the commands starting with \var are provided by $\mathcal{A}_{\mathcal{M}}\mathcal{S}$-LATEX.

Standard commands

α	\alpha	ι	\iota	τ	\tau
β	\beta	κ	\kappa	υ	\upsilon
γ	\gamma	λ	\lambda	ϕ	\phi
δ	\delta	μ	\mu	χ	\chi
ϵ	\epsilon	ν	\nu	ρ	\rho
ζ	\zeta	ξ	\xi	ψ	\psi
η	\eta	$\o$	\o	σ	\sigma
θ	\theta	π	\pi	ω	\omega

$\mathcal{A}_{\mathcal{M}}\mathcal{S}$-LATEX provided commands

ε	\varepsilon	$\varkappa$	\varkappa	ϱ	\varrho
φ	\varphi	ϑ	\vartheta	ϖ	\varpi
ς	\varsigma				

$\digamma$	\digamma

There are three classes of lowercase Greek letters. The following are the classes and the commands to typeset the letters in the classes.

regular symbols These are the regular lowercase Greek letters. The commands for typesetting these letters are \alpha (α), \beta (β), \gamma (γ), ...

additional italic symbols There are also some commands for additional or variant forms of italic lowercase Greek letters. These commands all start with \var: \varepsilon (ε), \vartheta (ϑ), \varrho (ϱ), ... These commands are provided by amsmath.

old number symbols Finally there is the $\mathcal{A}_{\mathcal{M}}\mathcal{S}$-LATEX-provided command \digamma, which gives you $\digamma$. The $\digamma$ symbol is the old Greek symbol for the number 6 [Bringhurst 2008, page 297].

There are also commands for uppercase Greek letters, but commands are only provided for letters that differ from the uppercase roman letters. For example, we don't need commands for uppercase letters such as A, B, E, and so on. There are two uppercase letter classes:

regular \Gamma (Γ), \Delta (Δ), \Theta (Θ), ... These are standard LATEX commands.

italic \varGamma ($\varGamma$), \varDelta ($\varDelta$), \varTheta ($\varTheta$), ... These non-standard commands produce italic-shaped uppercase Greek glyphs. These commands are provided by $\mathcal{A}_{\mathcal{M}}\mathcal{S}$-LATEX.

Table 8.1 lists the commands for the lowercase Greek letters. Table 8.2 lists the commands for the uppercase Greek letters. The commands at the top of both tables are standard. The remaining commands are provided by $\mathcal{A}_{\mathcal{M}}\mathcal{S}$-LATEX.

			Standard commands
Γ \Gamma	Ξ \Xi	Φ \Phi	
Δ \Delta	Π \Pi	Ψ \Psi	
Θ \Theta	Σ \Sigma	Ω \Omega	
Λ \Lambda	Υ \Upsilon		
			$\mathcal{AMS}$-LaTeX provided commands
$\varGamma$ \varGamma	$\varXi$ \varXi	$\varPhi$ \varPhi	
$\varDelta$ \varDelta	$\varPi$ \varPi	$\varPsi$ \varPsi	
$\varTheta$ \varTheta	$\varSigma$ \varSigma	$\varOmega$ \varOmega	
$\varLambda$ \varLambda	$\varUpsilon$ \varUpsilon		

Table 8.2
This table lists the math mode commands for uppercase Greek letters. The commands at the top of the table are standard LaTeX commands. The commands starting with \var are provided by amsmath.

8.6 Display Math Mode

This entire section is dedicated to display math material. Standard LaTeX provides a few commands for display math. The amsmath package redefines some of these commands and provides several extensions. As usual unstarred versions of the environments are numbered in the text. Starred versions are not numbered.

Some environments allow alignment positions in multi-line expressions.

○ The alignment positions are specified with &.
○ Line breaks are specified with \\.

The unstarred versions of the environment produce labels: equation, align, ... The starred versions of the environment do not produce labels: equation*, align*, ...

As a note of advice, you should avoid the unstarred versions unless you reference the equations in the text. If you decide otherwise, it may happen that a reader notices an unreferenced equation number and starts looking for the text that references the equation. Of course they cannot find the text location. (After several attempts!) They get confused and irritated. If the reader is a referee, this confusion and irritation may be just what they needed to reject your paper.

The remainder of this section consists of examples of some of amsmath's displayed equation environments. All examples use the unstarred versions.

8.6.1 *The equation Environment*

The equation environment is for typesetting a *single* numbered displayed equation. It is one of the most commonly used environments for typesetting display math material.

Figure 8.1 shows how to use the equation environment. The first thing to notice is that the display in the output makes the equation

```
... Binomial Theorem:
\begin{equation}
  \label{eq:Binomial}
  \sum^{n}_{i=0}
  \binom{n}{i}a^{i}b^{n-i}
    = (a+b)^{n}\,.
\end{equation}
Substituting $1$ for~$a$
 and $1$ for~$b$
 in~(\ref{eq:Binomial}) ...
```

The following is the Binomial Theorem:

$$\sum_{i=0}^{n}\binom{n}{i}a^ib^{n-i} = (a+b)^n. \quad (8.1)$$

Substituting 1 for a and 1 for b in (8.1) gives us $\sum_{i=0}^{n}\binom{n}{i} = 2^n$.

stand out clearly from the surrounding text. This is the main purpose of the display.

The example in Figure 8.1 uses a few new commands, which are explained as follows. The thin space command (\,) inserts a thin space just before the final punctuation mark. This makes the punctuation symbol stand out a little bit more and helps detecting it. The \sum command is for typesetting sums. The subscript (_) and superscript (^) commands are used to specify the lower and upper limits of the index variable in the summand. The superscript command also typesets the superscripts. The command \sum is explained in detail in Section 8.10. The command \binom typesets a binomial coefficient. Also notice that the equation in the output is automatically numbered and that the labelling and referencing mechanism in the input is standard:

○ The \label defines a label for an equation (number).
○ Applying the \ref command to an existing equation label results in the number of the equation. In Figure 8.1 we put the equation number inside parentheses: (\ref{eq:Binomial}). This is a common way to reference equations. Arguably it is better to use the prettyref package when typesetting references. This is explained in Section 1.5. You may also use the \eqref command, which automatically adds the parentheses for mathematical equations.

There is also a starred version (equation*) of the equation environment. As expected the environment equation* results in an unnumbered version of the equation environment. LaTeX also has a different mechanism for typesetting a single unnumbered equation. The command \[starts an unnumbered equation and the command \] ends it.

8.6.2 *The split Environment*

The split environment is for splitting a *single* equation into several lines. The environment provides support to align the resulting lines. The environment cannot be used at the top level and can only be used as part of other amsmath environments such as equation and gather. The split environment does not number the resulting equation. Fig-

```
\begin{equation*}
\begin{split}
  a & = b + c + d        \\
    & \qquad + f + g + h \\
    & > 0\,.
\end{split}
\end{equation*}
```

$$a = b + c + d$$
$$+ f + g + h$$
$$> 0.$$

Figure 8.2

The split environment

```
\begin{gather}
a = b\,,                  \\
\begin{split}
  a & = m + n + o        \\
    & \qquad + x + y + z \,.
\end{split}
\end{gather}
```

$$a = b, \qquad (8.2)$$
$$a = m + n + o$$
$$+ x + y + z. \qquad (8.3)$$

Figure 8.3

The gather environment

ure 8.2 shows how to use the environment. As you can see from the input and the resulting output, the alignment positions are indicated with the alignment tab token (&) in the input and linebreaks are forced with the newline operator (\\). This is the default mechanism for specifying alignment and linebreaks. The alignment position in the input in Figure 8.2 is just to the left of the equality symbol. This is why the line starting with a plus is indented a bit. This is done with the command \qquad, which is equivalent to two em spaces. Section 9.10.1 provides more information about the command \qquad.

As mentioned before, split does not number its output. This explains why there is no starred version of split.

Notice that there is no number in the output in Figure 8.2 because the equation* environment doesn't produce a number. However, we would have got one equation number if we had used equation instead of equation*.

8.6.3 *The gather Environment*

The gather environment displays a group of consecutive equations. All resulting equations are numbered and centred. The environment also has a starred version.

Figure 8.3 shows how to use the gather environment. Notice that the equation that is constructed using the split environment occupies two lines in the resulting output but only receives one number.

8.6.4 *The align Environment*

The align environment typesets equation groups with rows with horizontal alignment positions. Each row is numbered separately. The commands \nonumber and \notag turn off the numbering of the current equation.

152 | CHAPTER 8

Figure 8.4

Using the `align` environment

```
\begin{align}
   \label{eq:one}
   F( z ) & = \sum^\infty_{n=0} f_n z^n              \\
   \label{eq:two}
     & = z + \sum^\infty_{n=2} (f_{n-1}+f_{n-2}) z^n \\
   \label{eq:three}
     & = z + F( z )/z + F( z )/z^2                   \\
   \nonumber
     & = z / (1 - z - z^2)                           \,.
\end{align}
Here the last equation is obtained from~(\ref{eq:one}),
(\ref{eq:two}), and~(\ref{eq:three}) by transitivity
of equality and by solving for~$F( z )$.
```

Figure 8.5

Output of input in Figure 8.4

$$F(z) = \sum_{n=0}^{\infty} f_n z^n \tag{8.4}$$

$$= z + \sum_{n=2}^{\infty} (f_{n-1} + f_{n-2}) z^n \tag{8.5}$$

$$= z + F(z)/z + F(z)/z^2 \tag{8.6}$$

$$= z/(1 - z - z^2).$$

Here the last equation is obtained from (8.4), (8.5), and (8.6) by transitivity of equality and by solving for $F(z)$.

There is also a starred version of the `align` environment. Figure 8.4 shows how to use the starred version of the environment. The output is shown in Figure 8.5. The input in Figure 8.4 uses the command `\infty` to typeset the infinity symbol (∞). Notice that the `\nonumber` in the input suppresses the number of the corresponding equation/row in the output. The labels of the remaining three equations are defined as usual and are used in the text following the display to reference the equations.

The `align` environment may have several columns. Figure 8.6 and Figure 8.7 show how to do this.

Intermezzo.[Increasing productivity] Uniformity in the input format makes it easier to relate the input and the output. It also helps spotting inconsistencies, thereby reducing the possibility of errors in the input. Finally, it helps with debugging. For example, when you're creating complex output with the `align` environment it helps to have one (or a few) aligned items per input line. If you get an error with a given input then you can easily comment out these lines one by one until the error is gone. When the error disappears, this tells you the original error is in the vicinity of the last line you commented out. If you define multiple equations on a single input line then finding the error may pose more problems.

```
\begin{align}
  a_0 & = b_0\,, & b_0 & = c_0\,, & c_0 & = d_0\,,\\
  a_1 & = b_1\,, & b_1 & = c_1\,, & c_1 & = d_1\,,\\
  a_2 & = b_2\,, & b_2 & = c_2\,, & c_2 & = d_2\,.
\end{align}
```

Figure 8.6

The align environment

$$a_0 = b_0\,, \qquad b_0 = c_0\,, \qquad c_0 = d_0\,, \qquad (8.7)$$
$$a_1 = b_1\,, \qquad b_1 = c_1\,, \qquad c_1 = d_1\,, \qquad (8.8)$$
$$a_2 = b_2\,, \qquad b_2 = c_2\,, \qquad c_2 = d_2\,. \qquad (8.9)$$

Figure 8.7

Output of input in Figure 8.6

Figure 8.8

Using the \shortintertext command

```
\begin{align*}
  x_0 & = 0\,, \\
  x_1 & = 1\,, \\
\shortintertext{and}
  x_2 & = 2\,.
\end{align*}
```

$$x_0 = 0\,,$$
$$x_1 = 1\,,$$
$$\text{and}$$
$$x_2 = 2\,.$$

8.6.5 Interrupting a Display

The amsmath package also provides a command called \intertext for a short interjection of one or more lines in the middle of a multi-line display. The command \shortintertext, which is provided by the mathtools package, has a similar purpose but it takes less space. Figure 8.8 shows how you use it. (The command \intertext works in a similar way.) Notice that all equation symbols are properly aligned.

8.6.6 Low-level Alignment Building Blocks

All alignment environments we've seen so far operate at the "line" level. This means you cannot use them as parts of other constructs. The environments aligned, alignedat, and gathered align things at a lower level. Figure 8.9 provides an example of how to use the aligned environment. Notice that this environment does not do any number-ing: the numbering is controlled by the enclosing environment. In the input in Figure 8.9 the commands \left and \right scale the left and right square brackets, which act as delimiters of the construct that is built using aligned. The commands \left and \right are properly explained in Section 8.8.1. The example in Figure 8.9 should not be used as a general idiom for typesetting matrices because there are better techniques to typeset matrices. Section 9.3 explains this in further detail. More information about the other low-level alignment commands may be found in the amsmath documentation [AMS 2002].

Figure 8.9
The aligned environment

```
\begin{equation*}
   I = \left[
       \begin{aligned}
          1 && 0 && 0 \\
          0 && 1 && 0 \\
          0 && 0 && 1
       \end{aligned}
       \right]\,.
\end{equation*}
```

$$I = \begin{bmatrix} 1 & 0 & 0 \\ 0 & 1 & 0 \\ 0 & 0 & 1 \end{bmatrix}.$$

8.6.7 *The eqnarray Environment*

Standard LaTeX also has an eqnarray environment. This environment is traditionally used for multiple equations with one horizontal alignment position per line. The output from this environment it is not always satisfactory. TeXperts strongly recommend that you use the amsmath alignment environments instead.

8.7 Text in Formulae

Every now and then you need plain text in mathematical formulae. The amsmath package provides a command \text which lets you do this. Using it, is easy, as is demonstrated by the following example.

```
\[\text{final grade} =
  \text{\textsc{ca}} +
5 \times \text{exam}\,. \]
```

final grade = CA + 5 × exam.

Inside the argument of the \text command you can safely switch to ordinary math mode and back. You may also use \text in math mode in the argument of a \text command. This makes writing $f(\text{f($\text{f(f)})})$ perfectly valid but not particularly meaningful.

8.8 Delimiters

This section studies delimiters such as parentheses, which occur naturally in mathematical expressions. For example, the opening and closing parentheses act as delimiters of the start and end of the argument list of a function: $f(a)$, $g(x, y)$, and so on. Likewise, the symbol $|$ is used as a left and right delimiter in the commonly used notation $|\cdot|$ for absolute values. Despite the importance of delimiters, LaTeX is not always aware of their purpose and rôle in expressions. As a result LaTeX may sometimes use the wrong size and spacing in expressions with delimiters.

The remainder of this section helps you typeset your delimiters in the right size and with the correct spacing.

8.8.1 Scaling Left and Right Delimiters

We've already seen the commands \left and \right as part of an example, but this section properly describes the purpose of these commands. The main purpose of the commands \left and \right is to typeset *variable-sized* delimiters in the proper size.

To understand why we sometimes need to scale delimiters, consider the (artificial) LATEX expression $f(2^{2^{2^2}}_{2_{2_2}})$. If we typeset it using LATEX this gives us

$$f(2^{2^{2^2}}_{2_{2_2}}).$$

The resulting output is not very pretty because the parentheses, which act as delimiters of the arguments of $f(\cdot)$, are too small. LATEX is simply not aware that the parentheses are delimiters. To tell LATEX that the parentheses are left and right delimiters we make their purpose explicit by tagging them with \left and \right. This is done by writing $f\left(2^{2^{2^2}}_{2_{2_2}} \right)$, which gives us

$$f\left(2^{2^{2^2}}_{2_{2_2}}\right).$$

You can use this technique for any kind of variable-sized delimiter symbol. Section 8.8.5 presents the variable-sized delimiters.

You cannot use \left without \right and vice versa, which sometimes poses a problem. For example, how to typeset the following?

$$n! = \begin{cases} 1 & \text{if } 0 \leq n \leq 1, \\ n \times (n-1)! & \text{otherwise}. \end{cases}$$ LATEX Output

The following is the solution. In the solution we use a \right., which balances the \left-\right pair and produces nothing. The construct \left. may be used similarly.

```
\[ n! = \left\{
  \begin{aligned}
    & 1                && \text{if $0 \leq n \leq 1$}\,,
    \\ & n \times (n-1)! && \text{otherwise}\,.
  \end{aligned}
\right. \]
```
LATEX Input

Notice that the \{ in \left\{ in the input is not the left brace for starting a group, but the command for typesetting the left brace. The cases environment provides an easier way to define case-based definitions. This environment is explained in Section 9.6, which also discusses other solutions to case-based definitions.

When dealing with nested expressions and function applications, it is not always necessary to use parentheses. Sometimes your formulae may be clearer if you use square brackets at the outer level. Using square brackets at the outer level should work especially well in ordinary math mode because they avoid the need for scaling the delimiters.

```
Simplifying
$[ (a + b)^{2}
  - (a - b)^{2} ]^{2}$
gives us $16 a^{2} b^{2}$.
```

Simplifying $[(a+b)^2-(a-b)^2]^2$ gives us $16a^2b^2$.

A common problem is that of scaling opening and closing delimiters on different lines. The main problem is that each \left should be closed on the same line by a \right. Still there's a clever way out of this problem. The solution involves the \vphantom command. The idea is that you use the command to properly scale the smaller delimiter to the desired height in combination with \left or with \right. Since \vphantom{⟨stuff⟩} results in a box of zero width and the same height as ⟨stuff⟩, you can get a properly scaled delimiter if ⟨stuff⟩ is sufficiently high. A good candidate for ⟨stuff⟩ is the sub-formula on the line with the higher delimiter. Of course you should omit the delimiter. Other choices are also possible. The following is an example.

```
\begin{align*}
f & = g\left(3^{3^{3}}
    + ...\right.
\\& \qquad \left.
    + 3 \vphantom{3^{3^{3}}}
    \right)\,.
\end{align*}
```

$$f = g\left(3^{3^3} + ... \atop +3\right).$$

8.8.2 Bars

$\mathcal{AMS}$-LaTeX provides several commands for typesetting vertical bars (|). The reason for having several commands is that LaTeX's command \vert sometimes acts as a left, sometimes acts as a right delimiter, and sometimes acts as a different kind of delimiter. Depending on the rôle of the delimiter symbol it has to be treated differently, which is why $\mathcal{AMS}$-LaTeX provides dedicated commands that make the rôle of the delimiters explicit.

In the remainder of this section we shall first study the standard command \vert and then the special-purpose commands. The following demonstrates how to typeset the vertical bar in *guarded* sets.[1]

```
The even digits are
 given by
$\{\, 2 n \in \mathbb{N}
 \,\vert\,
     0 \leq n \leq 4 \,\}$.
```

The even digits are given by $\{2n \in \mathbb{N} \mid 0 \leq n \leq 4\}$.

Using the thin spaces before and after the vertical bar is slightly better than omitting them. It may be argued that using a colon in guarded set notations is better than using a bar. For example, despite the fact

---

[1] The adjective guarded for sets is inspired by guarded lists in the Haskell programming language [Peyton Jones, and Hughes 1999].

```
Let $F( z )$ be the ordinary
 generating function of
$\left\langle t_0,t_1,
 \ldots \right\rangle$. Then
$z F( z )$ is the ordinary
 generating function of
$\left\langle 0,t_0,t_1,
 \ldots \right\rangle$.
```

Let $F(z)$ be the ordinary generating function of $\langle t_0, t_1, \ldots \rangle$. Then $zF(z)$ is the ordinary generating function of $\langle 0, t_0, t_1, \ldots \rangle$.

Figure 8.10

Angular delimiters

that $\{|i| \mid -10 \le i \le 9\}$ doesn't involve complicated expressions, it is much more difficult to parse than $\{|i| : -10 \le i \le 9\}$. However, the bar notation also has its merits. For example, a bar is better in the following:

$$\{X : A \to B \mid Y : C \to A, X \circ Y = Z\}.$$

There are two more command-pairs for variable-size bars.

`$\left\lvert x \right\rvert$`

These commands are for absolute value-like expressions: $|x|$. ☑

`$\left\lVert x \right\rVert$`

These commands are for norms: $\|x\|$. ☑

The \rvert command is also used for the "evaluation at" notation. The following is an example. Notice that the \left. balances the \left-\right pair but typesets nothing.

```
\[\left. f( x )
  \right \rvert_{x=0} = 0\,.
\]
```

$$f(x)\Big|_{x=0} = 0.$$

8.8.3 Tuples

A common error in computer science and mathematical papers is to use $<1,2,3>$ for the tuple/sequence consisting of 1, then 2, and then 3. This kind of LaTeX input gives you $< 1, 2, 3 >$, which looks so bad that some authors have complained [Aslaksen 1993]. LaTeX has a special \langle and \rangle for tuples. If you use them for tuples then the result will look much more aesthetically pleasing. Figure 8.10 provides an example.

8.8.4 Floors and Ceilings

The commands \lfloor and \rfloor are for typesetting "floor" expressions, which are used to express rounding down. The two related commands \lceil and \rceil are for typesetting "ceiling" expressions. They are for rounding up.

Table 8.3

This table lists variable-size delimiters and the commands to typeset them. All delimiters are typeset in ordinary math mode. The delimiters listed under the heading 'Standard' are standard LaTeX-provided commands. The delimiters listed under the heading amsmath are provided by $\mathcal{A}\mathcal{M}\mathcal{S}$-LaTeX. All commands can be used with or without \left and \right.

					Standard
{	\{	}	\}	⟨	\langle
⌈	\lceil	⌊	\lfloor	⟩	\rangle
⌉	\rceil	⌋	\rfloor	↑	\uparrow
⇓	\Downarrow	↕	\updownarrow	↓	\downarrow
⇑	\Uparrow	⇕	\Updownarrow	(	(
[	[	|	|	)	)
]	]	‖	|	/	/
\	\backslash				

					amsmath
|	\lvert	|	\rvert		
‖	\lVert	‖	\rVert		

```
Let $x$ be any real number.
By definition
$i \leq
\left\lfloor x \right\rfloor
\leq x \leq
\left\lceil x \right\rceil
\leq I$
for all integers
$i$ and $I$ such that
$i \leq x \leq I$.
```

Let x be any real number. By definition $i \leq \lfloor x \rfloor \leq x \leq \lceil x \rceil \leq I$ for all integers i and I such that $i \leq x \leq I$.

8.8.5 Delimiter Commands

This section presents some more commands for variable-sized delimiters. Table 8.3 lists the commands. This table is based on [Pakin 2005, Tables 74 and 76].

8.9 Fractions

This section is about typesetting fractions in math mode. Ordinary fractions are typeset using the command \frac. To get the fraction $\frac{\langle\mathrm{num}\rangle}{\langle\mathrm{den}\rangle}$ you use \frac{⟨num⟩}{⟨den⟩}. Notice that fractions in the running text may disturb the flow of reading because they may increase the interline spacing. When using the \frac command in ordinary math mode you should ensure that the resulting interline spacing is acceptable. If it is not then perhaps it is possible to eliminate the division from your fractions. For example, a simple equation of the form $x = \frac{1}{3}y$ is equivalent to $3x = y$. If you can't eliminate division then perhaps you can turn the \frac construct into a simple $\langle\mathrm{num}\rangle/\langle\mathrm{den}\rangle$ construct. Alternatively, you can typeset the fraction in a display.

The amsmath package provides a specialised command \cfrac for typesetting continued fractions. The command takes an optional argument for the placement of the numerator. The value of this optional argument may be either l for left placement or r for right placement: you may write \cfrac[l]{⟨num⟩}{⟨den⟩} or \cfrac[r]{⟨num⟩}{⟨den⟩}. The following provides an example of how to use the command. In the example, the command \dotsb is for ellipsis in combination with binary relations or binary operators such as +.

```
\[ \sqrt{2} - 1
 = \cfrac{1}{2 +
   \cfrac{1}{2 +
   \dotsb}}}\,. \]
```

$$\sqrt{2} - 1 = \cfrac{1}{2 + \cfrac{1}{2 + \cdots}}.$$

8.10 Sums, Products, and Friends

This section describes how to typeset sums, products, integrals, and related constructs. Section 8.10.1 explains the basic typesetting commands. Section 8.10.2 explains how to control the lower and upper limits of delimited sums, products, and so on. Section 8.10.3 explains how to create multi-line upper and lower limits.

8.10.1 *Basic Typesetting Commands*

This section explains how to typeset basic sums, products, and related constructs. To get started we shall study with sums.

The *undelimited* sum symbol, $\sum$, may be typeset in math mode using the command \sum. It cannot be used in text mode.

In the *delimited* version the summand is parameterised by an index ranging from a lower to an upper limit. The subscript (_) and superscript (^) operators define the lower and upper limits of these delimited sums. So $\sum^n_{i = 0} f(i)$ defines the delimited sum with summand $f(i)$ and lower and upper limits for the index variable i, which are given by 0 and n respectively. The notation $\sum_{i=0}^{n} f(i)$ is a shorthand for $f(0) + f(1) + f(2) + \cdots + f(n)$.

In the *generalised* summation notation [Graham, Knuth, and Patashnik 1989, page 22] the range of the index variable is defined as a condition, which is defined in the same position as the lower limit. Examples of this form are $\sum_{0 \leq k < n} 2^{-k}$ and

$$\sum_{0 \leq k \leq n \,,\, \text{odd } k} 2^k \,.$$

If you study how the last two sums in the previous paragraph are typeset then you may notice that the conditions are typeset in different positions (relative to the $\sum$ symbol). This is not a coincidence. Indeed, in a display the limits usually appear below and above the summation symbol. However, in ordinary math mode they are positioned to the lower and upper right of the summation symbol. For ordinary math mode this avoids annoying discrepancies in interline spacing.

Table 8.4

This table lists variable-sized symbols and the commands to typeset them. The output in this table results from typesetting the commands in ordinary math mode. The commands in the first four rows of the table are standard LaTeX commands. The commands in the last row of the table are provided by the amsmath package.

					Standard
$\sum$	`\sum`	$\int$	`\int`	$\bigcap$	`\bigcap`
$\prod$	`\prod`	$\oint$	`\oint`	$\bigcup$	`\bigcup`
$\bigoplus$	`\bigoplus`	$\bigsqcup$	`\bigsqcup`	$\bigwedge$	`\bigwedge`
$\bigotimes$	`\bigotimes`	$\coprod$	`\coprod`	$\bigvee$	`\bigvee`
$\bigodot$	`\bigodot`	$\biguplus$	`\biguplus`		

					$\mathcal{AMS}$-LaTeX
$\iint$	`\iint`	$\iiint$	`\iiint`	$\iiiint$	`\iiiint`
$\int\cdots\int$	`\idotsint`				

The following is one more example with delimited sums.

```
According to folklore
Gauss proved that
$\sum{n}_{i=0} i=n(n+1)/2$.
```

According to folklore Gauss proved that $\sum_{i=0}^{n} i = n(n+1)/2$.

The $\sum$ symbol is an example of a *variable-sized* symbol [Lamport 1994, page 44]. Table 8.4 lists variable-sized symbols and the commands to typeset them. All the commands in the table are used in exactly the same way as you use the command `\sum`. The top of the table is based on [Lamport 1994, Table 3.8]. The commands in the top of the table are standard LaTeX commands. The commands in the last two rows are provided by the amsmath package.

8.10.2 *Overriding Text and Display Style*

Sometimes it is useful to change the way a variable-sized symbol is typeset. For example, a delimited sum in the numerator of a displayed fraction may look better if its limits are positioned to the lower and upper right of the $\sum$ symbol. The declarations `\textstyle` and `\displaystyle` specify the style of the variable-sized symbols. The following contrived example shows how to use the declaration `\textstyle`. The declaration `\displaystyle` is used similarly.

```
\[ \textstyle
   \sum^\infty_{n=0}
       2^{-n} = 2\,. \]
```

$\sum_{n=0}^{\infty} 2^{-n} = 2.$

8.10.3 *Multi-line Limits*

The `\substack` command lets you construct multi-line limits. Figure 8.11 uses the `\substack` command with the command `\sum`. As you may see from the input and the output, the `\\` command specifies a newline within the stack. All layers in the stack are centred.

```
\[ \sum_{\substack{
        \text{$i$ odd}}
        \\0 \leq i\leq n}}
      \binom{n}{i}
  = 2^n -
    \sum_{\substack{
        \text{$i$ even}
        \\0\leq i\leq n}}
      \binom{n}{i}\,. \]
```

$$\sum_{\substack{i\ \text{odd}\\0\leq i\leq n}}\binom{n}{i} = 2^n - \sum_{\substack{i\ \text{even}\\0\leq i\leq n}}\binom{n}{i}.$$

Figure 8.11

The \substack command with centred lines

```
\[ \sum_{\begin{subarray}{l}
        i \text{ odd}\\
        0 \leq i \leq n
      \end{subarray}}
      \binom{n}{i}
  = 2^n -
    \sum_{\begin{subarray}{c}
        i \text{ even}\\
        0 \leq i \leq n
      \end{subarray}}\,.
      \binom{n}{i} \]
```

$$\sum_{\begin{subarray}{l}i\ \text{odd}\\0\leq i\leq n\end{subarray}}\binom{n}{i} = 2^n - \sum_{\begin{subarray}{c}i\ \text{even}\\0\leq i\leq n\end{subarray}}\binom{n}{i}.$$

Figure 8.12

The subarray environment with different alignments. The alignment is controlled by the argument of the subarray environment. The lines in the first subarray are aligned to the left. The lines in the last subarray are centred.

The subarray environment provides more control than the \substack command. The environment takes one additional parameter, which specifies the alignment of the layers in the stack. The extra parameter can be l for alignment to the left or c for alignment to the centre. Figure 8.12 demonstrates how the environment enforces different alignments in the lines in the lower limits of sums. It may not be clear from the example but the spaces before the odd and even in the output are caused by the spaces in the argument of the \text commands. These spaces are typeset as visible spaces (␣) in the example. They are *not* caused by the spaces before the \text commands.

8.11 Existing Functions and Operators

LaTeX provides a wide range of function and operator symbols. The default type style for typesetting "log-like" function is math-roman (\mathrm). Table 8.5 lists LaTeX's built-in log-like functions.

Some operators take subscripts and/or superscripts. They work as usual: the subscripts are specified with the subscript operator (_) and the superscripts with the superscript operator (^). Figure 8.13 demonstrates how to get the limit of the \lim command in the subscript position. Note that Figure 8.13 also works if we omit the braces that turn the second argument of the superscript operator into a group. Adding the braces makes the second argument stand out a bit.

The mod symbol is overloaded. It requires different spacing depending on the context. The amsmath package provides four com-

Table 8.5
Log-like functions

arccos	\arccos	dim	\dim	log	\log
arcsin	\arcsin	exp	\exp	max	\max
arctan	\arctan	gcd	\gcd	min	\min
arg	\arg	hom	\hom	Pr	\Pr
cos	\cos	inf	\inf	sec	\sec
cosh	\cosh	ker	\ker	sin	\sin
cot	\cot	lg	\lg	sinh	\sinh
coth	\coth	lim	\lim	sup	\sup
csc	\csc	lim inf	\liminf	tan	\tan
deg	\deg	lim sup	\limsup	tanh	\tanh
det	\det	ln	\ln		

Figure 8.13
Limit of a log-like function

```
\[ \lim_{x \to 0}
        \frac{x^{2}}
            {x} = 0\,. \]
```

$$\lim_{x \to 0} \frac{x^2}{x} = 0.$$

mands to resolve this problem. The names of the commands are
\bmod, \mod, \pmod, and \pod. They are used as follows.

\bmod

This is for binary modular division: $\gcd(5, 3) = \gcd(3, 5 \bmod 3)$, which gives you $\gcd(5, 3) = \gcd(3, 5 \bmod 3)$. ☑

\mod

This is for modular equivalence: $2 \equiv 5 \mod 3$, which gives you $2 \equiv 5 \mod 3$. Notice the difference in spacing compared to the spacing you get with the command \bmod. Here the operator symbol, mod, is further to the right of its first argument. ☑

\pmod

This is for parenthesised modular equivalence: $2 \equiv 5 \pmod 3$, which gives you $2 \equiv 5 \pmod 3$. ☑

\pod

This is for parenthesised modular equivalence without mod symbol: $2 \equiv 5 \pod 3$, which gives you $2 \equiv 5 \; (3)$. ☑

8.12 Integration and Differentiation

8.12.1 Integration

The command \int is for typesetting simple integrals. There are three styles in the literature for typesetting integrals. The first style is mainly used by mathematicians; the other styles are used in engineering and physics [Beccari 1997].

The following shows how mathematicians typeset a definite integral. This style may be found for example in [Clapham, and Nicholson 2005]. Notice that the standard \left.-\right\rvert-trick scales the \rvert to the correct size. The thin space is needed for each variable of integration (the dx), so in general you write \,d x\,d y, and so on.

Table 8.6

This table lists integration signs and the commands to typeset them. All commands except \int are provided by the amsmath package. The first five commands are provided by the amsmath package. The remaining commands are provided by the esint package.

			amsmath
$\int$	\int	$\iint$	\iint
$\iiint$	\iiint	$\iiiint$	\iiiint
$\int\cdots\int$	\idotsint		
			esint
$\int$	\int	$\iint$	\iint
$\iiint$	\iiintop	$\iiiint$	\iiiintop
$\oiint$	\sqint	$\oiiint$	\sqiint
$\ointctrclockwise$	\ointctrclockwise	$\ointclockwise$	\ointclockwise
$\landupint$	\landupint	$\landdownint$	\landdownint
$\fint$	\fint	$\int\cdots\int$	\dotsintop
$\ointop$	\ointop	$\oiintop$	\oiintop
$\varointctrclockwise$	\varointctrclockwise	$\varointclockwise$	\varointclockwise
$\varoiint$	\varoiint		

```
\[ \int^{b}_{a}
   3 x^{2}\,d x
= \left. x^{3}
   \right\rvert^{b}_{a}
= b^{3} - a^{3}\,. \]
```

$$\int_a^b 3x^2\,dx = x^3\Big|_a^b = b^3 - a^3\,.$$

There are two more alternative styles for typesetting integrals. They differ in how they treat the variable of integration. Both styles typeset the d in an upright (roman) typeface, which is common in physics and engineering [Beccari 1997].

The first alternative style omits the thin space and uses the \mathrm for the variables of integration: \mathrm{d x}. This style may be found in [Borowski, and Borwein 2005]:

$$\int_a^b 3x^2\mathrm{d}x\,.$$

The second alternative style may be found in [Zeidler 1996]. It uses a mixture of the other styles:

$$\int_a^b 3x^2\mathrm{d}x\,.$$

From now on we shall use the first style when it comes to typesetting the variable of integration/differentiation.

The key to typesetting other integrals are the commands provided by the amsmath and the esint packages. Table 8.6 lists these commands.

8.12.2 *Differentiation*

Expressions with differentiations are typeset using the \frac command. The expression $\frac{du}{dx}$ may be obtained with \frac{d u}{d x}. More complex expressions work as expected.

```
\[ \frac{d^{2} u}
     {d x^{2}} \]
```

$$\frac{d^2 u}{dx^2}$$

The symbol ∂ is typeset with the command \partial. Partial derivatives with ∂ are created with fractions:

```
Let $z = x^{2} + x y$, then
\[ \frac{\partial z}
       {\partial x}
   = 2x + y\,. \]
```

Let $z = x^2 + xy$, then
$$\frac{\partial z}{\partial x} = 2x + y.$$

8.13 Roots

Square roots and other roots are typeset with \sqrt. The command has an optional argument for the root indices.

```
$\sqrt{2} \approx
   1.414213562$ and
$\sqrt[3]{27} = 3$.
```

$\sqrt{2} \approx 1.414213562$ and $\sqrt[3]{27} = 3$.

The placement of the root indices is not always perfect. The amsmath package provides two commands to fine-tune the root index placement:

\leftroot{⟨number⟩}
This moves the root index ⟨number⟩ "units" to left. The unit is an arbitrary but convenient distance. Notice that ⟨number⟩ can be negative, in which case this results in moving the root index to the right. ☑

\uproot{⟨number⟩}
This moves the root index ⟨number⟩ units up. ☑

The commands are used in the optional argument. The following examples shows how to improve the poor positioning of the root index β that you get with $\sqrt[\beta]{k}$.

```
We all agree that
$\sqrt[\beta]{k}$
 is equal to
$\sqrt[\leftroot{-2}%
     \uproot{2}%
     \beta]{k}$.
But why are they
 different in type?
```

We all agree that $\sqrt[\beta]{k}$ is equal to $\sqrt[\beta]{k}$. But why are they different in type?

8.14 Changing the Style

The following six commands change the type style in math mode.

$\mathit{italic + abc^2}$
The command \mathit typesets its argument in the math italic typeface: $\mathit{italic + abc^2}$. ☑

⨿	`\amalg`	◇	`\diamond`	⊓	`\sqcap`
∗	`\ast`	÷	`\div`	⊔	`\sqcup`
◯	`\bigcirc`	◁	`\lhd`	⋆	`\star`
▽	`\bigtriangledown`	∓	`\mp`	×	`\times`
△	`\bigtriangleup`	⊙	`\odot`	◁	`\triangleleft`
●	`\bullet`	⊖	`\ominus`	▷	`\triangleright`
∩	`\cap`	⊕	`\oplus`	⊴	`\unlhd`
·	`\cdot`	⊘	`\oslash`	⊵	`\unrhd`
∘	`\circ`	⊗	`\otimes`	⊎	`\uplus`
∪	`\cup`	±	`\pm`	∨	`\vee`
†	`\dagger`	▷	`\rhd`	∧	`\wedge`
‡	`\ddagger`	\	`\setminus`	≀	`\wr`

Table 8.7

Binary operation symbols. The commands `\lhd`, `\rhd`, `\unlhd`, and `\unrhd` are provided by the `amssymb` package.

`$\mathrm{roman + abc^2}$`

The command `\mathrm` typesets its argument in math roman typeface: $\mathrm{roman} + abc^2$. ☑

`$\mathbf{bold + abc^2}$`

The command `\mathbf` typesets its argument in the default math bold typeface: $\mathbf{bold} + \mathbf{abc^2}$. Notice that `\mathbf` may not always result in bold symbols. Although not ideal, the commands `\pmb` (poor man's bold) and `\boldsymbol` may be useful in cases like this. ☑

`$\mathsf{sans serif + abc^2}$`

The command `\mathsf` typesets its argument in the math sans serif typeface: $\mathsf{sansserif} + abc^2$. ☑

`$\mathtt{teletype + abc^2}$`

The command `\mathtt` typesets its argument in math monospaced typeface: $\mathtt{teletype} + abc^2$. ☑

`$\mathcal{CALLIGRAPHIC}$`

The command `\mathcal` typesets its argument in a math calligraphic typeface: $\mathcal{CALLIGRAPHIC}$. The calligraphic letters only come in uppercase: ☑

8.15 Symbol Tables

This section presents various tables with commands math mode symbols. The presentation is based on [Lamport 1994] and [Pakin 2005].

8.15.1 *Operator Symbols*

Table 8.7 lists all LaTeX-provided binary operator symbols.

8.15.2 *Relation Symbols*

LaTeX's list of relation symbols is quite impressive. Table 8.8 lists LaTeX's built-in symbols for binary relations. The `amsmath` package provides additional commands. Table 8.9 lists the commands.

Table 8.8

Relation symbols. The commands \Join, \sqsubset, and \sqsupset are provided by the amssymb package.

<	\<	=	\=	≤	\leq
>	\>	≪	\ll	⌣	\smile
≈	\approx	|	\mid	⊑	\sqsubseteq
≍	\asymp	⊨	\models	⊏	\sqsubset
⋈	\bowtie	≠	\neq	⊒	\sqsupseteq
≅	\cong	∋	\ni	⊐	\sqsupset
⊣	\dashv	∉	\notin	⊆	\subseteq
≐	\doteq	‖	\parallel	⊂	\subset
≡	\equiv	⊥	\perp	≽	\succeq
⌢	\frown	⪯	\preceq	≻	\succ
≥	\geq	≺	\prec	⊇	\supseteq
≫	\gg	∝	\propto	⊃	\supset
∈	\in	≃	\simeq	⊢	\vdash
⋈	\Join				
∼	\sim				

Table 8.9

Additional relational symbols. These symbols are provided by the amsmath package.

≊	\approxeq	≖	\eqcirc	⪸	\succapprox
϶	\backepsilon	≒	\fallingdotseq	≽	\succcurlyeq
∽	\backsim	⊸	\multimap	≿	\succsim
⋍	\backsimeq	⋔	\pitchfork	∴	\therefore
∵	\because	⪷	\precapprox	≈	\thickapprox
≬	\between	≼	\preccurlyeq	∼	\thicksim
≏	\Bumpeq	≾	\precsim	∝	\varpropto
≎	\bumpeq	≓	\risingdotseq	⊩	\Vdash
≗	\circeq	⥳	\shortmid	⊨	\vDash
⋞	\curlyeqprec	∥	\shortparallel	⊪	\Vvdash
⋟	\curlyeqsucc	⌢	\smallfrown	≑	\doteqdot
⌣	\smallsmile				

8.15.3 Arrows

LaTeX defines several commands for drawing arrows. All these commands produce fixed-size arrows. Extensible arrows are provided by additional packages. Table 8.10 lists all LaTeX's built-in commands for fixed-size arrows. Some commands for extensible arrows are listed in Tables 8.11–8.13. These commands, some of which accept an optional argument, require additional packages.

8.15.4 Miscellaneous Symbols

Table 8.14 lists LaTeX's "miscellaneous" symbols. It is worthwhile noticing that the commands \imath and \jmath produce a dotless ı and a dotless ȷ. These symbols should be used in combination with hats and similar decorations; never use i and j with hats.

↓	\downarrow	⇓	\Downarrow
↑	\uparrow	⇑	\Uparrow
↕	\updownarrow	⇕	\Updownarrow
←	\leftarrow	⇐	\Leftarrow
→	\rightarrow	⇒	\Rightarrow
⟵	\longleftarrow	⟸	\Longleftarrow
⟶	\longrightarrow	⟹	\Longrightarrow
↔	\leftrightarrow	⇔	\Leftrightarrow
⟷	\longleftrightarrow	⟺	\Longleftrightarrow
↦	\mapsto	↩	\hookleftarrow
⟼	\longmapsto	↪	\hookrightarrow
↼	\leftharpoonup	↗	\nearrow
↽	\leftharpoondown	↘	\searrow
⇀	\rightharpoonup	↙	\swarrow
⇁	\rightharpoondown	↖	\nwarrow
⇌	\rightleftharpoons		

Table 8.10

Fixed-size arrow symbols

$\xleftarrow{e}$	\xleftarrow{e}	$\xleftarrow[o]{e}$	\xleftarrow[o]{e}
$\xrightarrow{e}$	\xrightarrow{e}	$\xrightarrow[o]{e}$	\xrightarrow[o]{e}
$\underleftarrow{e}$	\underleftarrow{e}	$\underrightarrow{e}$	\underrightarrow{e}
$\overleftrightarrow{e}$	\overleftrightarrow{e}	$\underleftrightarrow{e}$	\underleftrightarrow{e}

Table 8.11

Extensible arrow symbols provided by amsmath

$\xleftrightharpoons{e}$	\xleftrightharpoons{e}	$\xrightleftharpoons{e}$	\xrightleftharpoons{e}
$\xleftharpoondown{e}$	\xleftharpoondown{e}	$\xrightharpoondown{e}$	\xrightharpoondown{e}
$\xleftharpoonup{e}$	\xleftharpoonup{e}	$\xrightharpoonup{e}$	\xrightharpoonup{e}
$\xleftrightarrow{e}$	\xleftrightarrow{e}	$\xLeftrightarrow{e}$	\xLeftrightarrow{e}
$\xhookleftarrow{e}$	\xhookleftarrow{e}	$\xhookrightarrow{e}$	\xhookrightarrow{e}
$\xLeftarrow{e}$	\xLeftarrow{e}	$\xRightarrow{e}$	\xRightarrow{e}
$\xmapsto{e}$	\xmapsto{e}		

Table 8.12

This table lists non-standard mathtools-provided extensible arrow symbols and the commands to typeset them. All these commands also take an optional argument. Table 8.13 demonstrates the effect of using the options.

```
Some people write
$\hat{i}$ and $\hat{j}$
but $\hat{\imath}$ and
$\hat{\jmath}$ is better.
```
Some people write $\hat{\imath}$ and $\hat{\jmath}$ but $\hat{\imath}$ and $\hat{\jmath}$ is better.

In some (math) typefaces it is difficult to distinguish the letter l (1) from the digit 1 (1). In such cases you may want to consider using the letter ℓ (ℓ) as a substitute for the letter l.

Table 8.13
This table lists non-standard mathtools-provided extensible arrow symbols and the commands to typeset them. Table 8.12 lists how these commands work without the optional argument.

$\overset{e}{\underset{o}{\leftrightharpoons}}$	`\xleftrightharpoons[o]{e}`	$\overset{e}{\underset{o}{\rightleftharpoons}}$	`\xrightleftharpoons[o]{e}`
$\overset{e}{\underset{o}{\leftharpoondown}}$	`\xleftharpoondown[o]{e}`	$\overset{e}{\underset{o}{\rightharpoondown}}$	`\xrightharpoondown[o]{e}`
$\overset{e}{\underset{o}{\leftharpoonup}}$	`\xleftharpoonup[o]{e}`	$\overset{e}{\underset{o}{\rightharpoonup}}$	`\xrightharpoonup[o]{e}`
$\overset{e}{\underset{o}{\leftrightarrow}}$	`\xleftrightarrow[o]{e}`	$\overset{e}{\underset{o}{\Leftrightarrow}}$	`\xLeftrightarrow[o]{e}`
$\overset{e}{\underset{o}{\hookleftarrow}}$	`\xhookleftarrow[o]{e}`	$\overset{e}{\underset{o}{\hookrightarrow}}$	`\xhookrightarrow[o]{e}`
$\overset{e}{\underset{o}{\Leftarrow}}$	`\xLeftarrow[o]{e}`	$\overset{e}{\underset{o}{\Rightarrow}}$	`\xRightarrow[o]{e}`
$\overset{e}{\underset{o}{\mapsto}}$	`\xmapsto[o]{e}`		

Table 8.14
Miscellaneous math mode symbols. The commands `\Box`, `\Diamond`, and `\mho` are provided by the amssymb package.

$\aleph$	`\aleph`	$\flat$	`\flat`	$\neg$	`\neg`
$\angle$	`\angle`	$\forall$	`\forall`	$\Re$	`\Re`
$\backslash$	`\backslash`	$\hbar$	`\hbar`	$\surd$	`\surd`
$\bot$	`\bot`	$\heartsuit$	`\heartsuit`	$\top$	`\top`
$\Box$	`\Box`	$\Im$	`\Im`	$\triangle$	`\triangle`
$\clubsuit$	`\clubsuit`	$\imath$	`\imath`	∂	`\partial`
$\Diamond$	`\Diamond`	∞	`\infty`	$\prime$	`\prime`
$\diamondsuit$	`\diamondsuit`	$\jmath$	`\jmath`	$\sharp$	`\sharp`
ℓ	`\ell`	$\mho$	`\mho`	$\spadesuit$	`\spadesuit`
$\emptyset$	`\emptyset`	∇	`\nabla`	$\wp$	`\wp`
$\exists$	`\exists`	$\natural$	`\natural`	$\|$	`|`

CHAPTER 9
Advanced Mathematics

THIS CHAPTER COVERS advanced mathematical typesetting and the related commands. If you're reading this book for the first time, you may want to skip the entire chapter.

9.1 Declaring New Operators

Every TeXnician some day runs out of operator or function symbols. Fortunately, the amsmath package provides a high-level command for defining user-defined operator commands. The resulting commands properly typeset their operators and functions in a consistent style. The command gives you full control over the positioning of subscripts and superscripts in "limit" positions. The command, which is called \DeclareMathOperator, can only be used in the preamble.

\DeclareMathOperator{⟨command⟩}{⟨sym⟩}

This defines a new command, ⟨command⟩, which is typeset as ⟨sym⟩. The ⟨command⟩ should start with a backslash (\). The resulting symbol is typeset in a uniform style and with the proper spacing. The following is an example. ☑

```
\documentclass{article}
\DeclareMathOperator\op{op}
\begin{document}
  ... Note that
  $1 \mathrm{op} 2 = 3$
  does not look pretty.
  However, $1 \op 2 = 3$
  looks good.
```

... Note that 1op2 = 3 does not look pretty. However, 1 op 2 = 3 looks good.

Notice that the appearance of both operator symbols in the previous example is identical. However, the spacing for the first operator symbol is dreadful because LaTeX does not recognise it as an operator.

$\mathcal{AMS}$-LaTeX also provides a \DeclareMathOperator* command, which is for defining operator symbols that require subscripts and superscripts in "limit" positions. It can only be used in the preamble. The following is an example.

We start by defining the operator symbol in the preamble. This is done as follows.

```
\DeclareMathOperator*\Lim{Lim}
```
LaTeX Input

Figure 9.1

The array environment

```
\[ \left(
   \begin{array}{c}
      \left\lvert
      \begin{array}{lrc}
         x & y & z
         \\ 2 a & 3 b & 4 c
      \end{array}
      \right\rvert
      \\ \alpha
      \\ \beta
   \end{array}
   \right) \]
```

$$\left(\left| \begin{array}{lrc} x & y & z \\ 2a & 3b & 4c \end{array} \right| \atop \begin{array}{c} \alpha \\ \beta \end{array} \right)$$

We continue by using the new operator command.

```
$\Lim_{x \to 0}
   \frac{x^{2}}{x} = 0$....
```
... $\Lim_{x\to 0} \frac{x^2}{x} = 0$. ...

9.2 Managing Content with the `cool` Package

The `cool` package addresses the problem of capturing content.

○ The package provides a *very* comprehensive list of commands for consistent typesetting of mathematical functions and constants.
○ The package also provides commands for easy typesetting complex matrices.
○ Finally, the package provides commands that determine how symbols and expressions are typeset. This affects:
  ⋆ How inverse trigonometric functions are typeset: $\arcsin x$ versus $\sin^{-1} x$.
  ⋆ How to typeset derivatives: $\frac{d}{dx}f$ versus $\frac{df}{dx}$.
  ⋆ How to print certain function and polynomial symbols.
  ⋆ How to typeset integrals: $\int f\, dx$, versus $\int f dx$, versus, $\int dx f$, ...

9.3 Arrays and Matrices

Traditionally arrays were typeset using the `array` environment, which works similar to the `tabbing` environment. Figure 9.1 provides an example that shows LaTeX's built-in `array` environment.

The `amsmath` package provides six environments for matrices. All commands are designed for display math mode. They don't provide alignment control. By default there are up to ten columns, which are aligned to the centre. The following are the commands.

`pmatrix` For matrices with parentheses as delimiters: $\begin{pmatrix} 1 & 2 & 3 \end{pmatrix}$.
`bmatrix` For matrices with square brackets as delimiters: $\begin{bmatrix} 1 & 2 & 3 \end{bmatrix}$.
`Bmatrix` For matrices with braces as delimiters: $\begin{Bmatrix} 1 & 2 & 3 \end{Bmatrix}$.
`vmatrix` For matrices with vertical bars as delimiters: $\begin{vmatrix} 1 & 2 & 3 \end{vmatrix}$.

```
... The linear transformation
$\langle\,x,y\,\rangle
 \mapsto
 \langle\,2x + y, y\,\rangle$
is written as follows:
$\bigl[\begin{smallmatrix}
   2&1 \\ 0&1
 \end{smallmatrix}\bigr]
 \bigl[\begin{smallmatrix}
   x   \\ y
 \end{smallmatrix}\bigr]$.
```

Figure 9.2

The smallmatrix environment

... The linear transformation $\langle x,y \rangle \mapsto \langle 2x+y,y \rangle$ is written as follows: $\left[\begin{smallmatrix} 2 & 1 \\ 0 & 1 \end{smallmatrix}\right]\left[\begin{smallmatrix} x \\ y \end{smallmatrix}\right]$.

Vmatrix For matrices with two double vertical bars as delimiters: $\left\| \begin{smallmatrix} 1 & 2 & 3 \end{smallmatrix} \right\|$.

matrix For matrices without delimiters: $\begin{smallmatrix} 1 & 2 & 3 \end{smallmatrix}$.

The following example shows how to use the bmatrix environment.

```
After rotation this gives
\[ \begin{bmatrix}
    \cos\phi & -\sin\phi
   \\\sin\phi & \cos\phi
   \end{bmatrix}
   \begin{bmatrix}
    x \\ y
   \end{bmatrix}\,. \]
```

After rotation this gives
$$\begin{bmatrix} \cos\phi & -\sin\phi \\ \sin\phi & \cos\phi \end{bmatrix}\begin{bmatrix} x \\ y \end{bmatrix}.$$

$\mathcal{A}_{\mathcal{M}}\mathcal{S}$-LaTeX also provides a smallmatrix environment for matrices in inline (ordinary) math mode. The smallmatrix environment does not typeset the delimiters. To typeset the delimiters you use the commands \bigl and \bigr, which are the equivalents of the commands \left and \right respectively, so for square bracket delimiters you write: $\bigl[\begin{smallmatrix} ... \end{smallmatrix}\bigr]$. Figure 9.2 provides an example.

9.4 Accents, Hats, and Other Decorations

This section is about typesetting accents and other decorations in math mode. The commands in this section are all of the form ⟨command⟩{⟨argument⟩}. The majority of the commands add a fixed-size decoration to the ⟨argument⟩. For example, $\hat{x}$ and $\hat{xxx}$ give you $\hat{x}$ and $x\hat{x}x$. The remaining commands provide extensible decorations. For example, the commands $\overline{x}$ and $\overline{xxx}$ give you $\overline{x}$ and $\overline{xxx}$. The result of some decorations may not always be aesthetically pleasing. For example, $\widetilde{xxxxxx}$ gives you $\widetilde{xxxxxx}$. As a different example, \bar{x} results in $\bar{x}$, which has a more subtle bar than \overline{x}, which gives you $\overline{x}$. Table 9.1 lists some commonly used commands.

Table 9.1

Math mode accents, hats, and other decorations. The commands at the top are listed with single letter arguments. They are intended for "narrow" arguments such as letters, digits, and so on. The commands at the bottom produce extensible decorations. The commands \dddot and \ddddot are provided by amsmath.

	Fixed-size Decorations	
$\dot{x}$ \dot{x}	$\acute{x}$ \acute{x}	
$\ddot{x}$ \ddot{x}	$\grave{x}$ \grave{x}	
$\dddot{x}$ \dddot{x}	$\hat{x}$ \hat{x}	
$\ddddot{x}$ \ddddot{x}	$\tilde{x}$ \tilde{x}	
$\mathring{x}$ \mathring{x}	$\bar{x}$ \bar{x}	
$\check{x}$ \check{x}	$\vec{x}$ \vec{x}	
$\breve{x}$ \breve{x}		

	Extensible Decorations	
$\overleftarrow{e}$ \overleftarrow{e}	$\overline{e}$ \overline{e}	
$\overrightarrow{e}$ \overrightarrow{e}	$\widetilde{e}$ \widetilde{e}	
$\overleftrightarrow{e}$ \overleftrightarrow{e}	$\widehat{e}$ \widehat{e}	
$\underleftarrow{e}$ \underleftarrow{e}	$\underbar{e}$ \underbar{e}	
$\underleftrightarrow{e}$ \underleftrightarrow{e}	$\underline{e}$ \underline{e}	
$\underrightarrow{e}$ \underrightarrow{e}		

Table 9.2

The \overbrace and \underbrace commands

$\overbrace{u}^{o}$ \overbrace{u}^{o}	$\overbrace{u}$ \overbrace{u}
$\underbrace{o}_{u}$ \underbrace{o}_{u}	$\underbrace{o}$ \underbrace{o}

9.5 Braces

Every now and then you may have to typeset expressions with overbraces ($\overbrace{\langle expr\rangle}$) and underbraces ($\underbrace{\langle expr\rangle}$). The commands for creating such expressions are \overbrace and \underbrace. As should be clear from the second line in this paragraph, overbraces and underbraces should be restricted to display math mode because they may affect the line spacing.

Expressions with underbraces can be "decorated" with expressions under the brace. Likewise, expressions with overbraces may receive decorations over the brace. These more complicated expressions are constructed using subscript and superscript operators. Table 9.2 demonstrates how to use the \underbrace and \overbrace commands.

The decorated versions are usually needed to indicate numbers of subterms. Figure 9.3 provides an example. (Notice the use of the \text command to temporarily switch to text mode inside math mode.)

9.6 Case-based Definitions

Case-based definitions are very common in computer science. There are two common approaches and solutions: conditions and *Iversonians*.

```
\[ x^{k} =
   \underbrace
      {1 \times x
       \times x \times
       \dotsb \times x}
      _{\text{$k$~times
               $\times x$}} \,. \]
```

$$x^k = \underbrace{1 \times x \times x \times \cdots \times x}_{k \text{ times } \times x}.$$

Figure 9.3
Typesetting an underbrace

conditions With this approach each condition results in a case in a case-based definition. The following provides an example.

LaTeX Input

```
\[ n! = \begin{cases}
        1                 & \text{if $n = 0$}\,; \\
        (n-1) ! \times n & \text{if $n > 0$}\,.
        \end{cases}  \]
```

This gives you the following.

LaTeX Output

$$n! = \begin{cases} 1 & \text{if } n = 0; \\ (n-1)! \times n & \text{if } n > 0. \end{cases}$$

Formulae like this look great in a display but aren't suitable for ordinary math mode.

Iversonians Here we define a unary *characteristic* or *indicator* function that returns 1 if its argument is true and returns 0 otherwise. Graham, Knuth, and Patashnik [1989] propose the notation [⟨cond⟩], which they call the *Iversonian* of ⟨cond⟩ as a tribute to Kenneth E. Iverson, the inventor of the computer language A Programming Language (APL), which has a similar construct. The expression evaluates to 1 if ⟨cond⟩ is true and 0 otherwise. The notation $1_{\{⟨cond⟩\}}$ is also accepted notation, but it has the disadvantage that it has a subscript, which may disturb the line spacing in ordinary math mode. The following example shows that Iversonians work surprisingly well in ordinary math mode.

```
... We define
$n! = [\,n = 0\,] +
  (n-1) ! \times n
  \times [\,n > 0\,]$. ...
```

... We define $n! = [n = 0] + (n-1)! \times n \times [n > 0]$. ...

9.7 Function Definitions

Function definitions usually come with a description of the domain, the range, and the computation rule.

```
The successor function,
$s \colon \mathbb{N}
    \to    \mathbb{N}$,
 is defined as follows:
\[ s( n ) \mapsto n+1 \,. \]
```

The successor function, $s: \mathbb{N} \to \mathbb{N}$, is defined as follows:
$$s(n) \mapsto n + 1.$$

Notice that \to and \mapsto result in different arrows. Also a colon (:) should not be used as a substitute for \colon because it gives you $s:\mathbb{N}\to\mathbb{N}$. The \mathbb command typesets it argument in a blackboard typeface. This is useful for typesetting the symbols $\mathbb{N}, \mathbb{Z}, \mathbb{Q}, \mathbb{R}, \mathbb{C}$, and related symbols. The command can only be used in math mode.

9.8 Theorems

The package amsthm makes writing theorems, lemmas, and friends easy. The package ensures consistent numbering and appearance of theorem-like environments. The package provides:

○ A proof environment;
○ Styles for theorem-like environments;
○ Commands for defining new theorem-like styles; and
○ Commands for defining new theorem-like environments.

9.8.1 *Theorem Taxonomy*

The following is the typical output of a theorem-style environment.

> **Theorem 2.1.3** (Fermat's Last Theorem). *Let n be any integer greater than 2, then the equation $a^n + b^n = c^n$ has no solutions in positive integers a, b, and c.*
>
> LaTeX Output

The definition consists of several parts.

heading The heading should describe the rôle of the environment. In this example the heading is 'Theorem.' Usually, headings are Theorem, Lemma, Definition, and so on.

number The number is optional and is used to reference the environment in the running text. This is done using the usual \label-\ref mechanism. Numbers may depend on sectional units. If the number depends on sectional units then it is of the form ⟨unit⟩.⟨number⟩, where ⟨unit⟩ is the number of the current sectional unit, and ⟨number⟩ is a number that is local within the sectional unit. If the number does not depend on the sectional unit then it is a plain number. In this example, the number of the environment is 2.1.3. This indicates that the number depends on the sectional unit 2.1—probably Chapter 2.1 or Section 2.1—and that within the unit this is the third instance.

body The body of the environment is the text that conveys the message of the environment.

name The name is optional and serves two purposes. Most importantly, the name should capture the essence of the theorem. Next, it may be used to reference the environment by name, as opposed to by number (using \ref). In this example, the theorem's name is *Fermat's Last Theorem*.

9.8.2 Styles for Theorem-like Environments

The `amsthm` package defines three styles for theorem-like environments: `plain`, `definition`, and `remark`. New styles may also be defined; this is explained in Section 9.8.4. The following explains the differences between the existing styles.

plain Usually associated with: Theorem, Lemma, Corollary, Proposition, Conjecture, Criterion, and Algorithm. The following demonstrates the appearance of the `plain` style.

> **Theorem 1.1** (Fermat's Last Theorem). *Let n be any integer greater than 2, then the equation $a^n + b^n = c^n$ has no solutions in positive integers a, b, and c.*   LATEX Output

definition Usually associated with: Definition, Condition, Problem, and Example. The following demonstrates the appearance of the `definition` style.

> **Definition 1.2** (Ceiling). The *ceiling* of real number, r, is the smallest integer, i, such that $r \leq i$.   LATEX Output

remark Usually associated with: Remark, Note, Notation, Claim, Summary, Acknowledgement, Case, and Conclusion. The following demonstrates the appearance of the `remark` style.

> **Tip 1.3** (Tip). Don't do this at home.   LATEX Output

Numbering may or may not depend on the sectional unit. The following explains the differences.

independent numbering Here the numbers are integers. So if theorems are numbered continuously you may have Theorem 1, Theorem 2, Theorem 3, and so on.

dependent numbering Here the numbers are of the form ⟨unit⟩.⟨local⟩, where ⟨unit⟩ depends on the sectional unit number, and ⟨local⟩ is a local number. With numbering dependent on a section in a book you may have Theorem 1.1.1, Theorem 1.1.2, Theorem 2.3.1, and so on.

Different environments may or may not share number sequences.

with sharing Environments may share the same number sequence. If this is the case you may get Theorem 1, Lemma 2, Theorem 3, and so on. However, you cannot have both Lemma 2 and Theorem 2.

without sharing Environments do not share their number sequence. Instead, they have their own independent number sequence. If this is the case you could get Theorem 1, Lemma 1, Theorem 2, and so on.

9.8.3 Defining Theorem-like Environments

Defining new theorem-like environment styles is done in two stages. First you set the current style, next you define the environments. The environments will all be typeset in the style that was current at the time of definition of the environments.

defining the current style Defining the current style is done with the \theoremstyle command. The command takes the label of the style as its argument. Initially, the current style is plain.

defining the next environment Defining the next environment is done with the \newtheorem command. The environment is typeset in the style that was current at the time of definition; this coincides with the call to \newtheorem. The numbering and heading of the environment is determined by the command \newtheorem, which takes an optional argument that may appear in different positions.

The remainder of this section explains the \newtheorem command. The command may be used with or without an optional argument.

without an optional argument Without the optional argument you define environments using \newtheorem{⟨env⟩}{⟨heading⟩}. This defines a new environment ⟨env⟩ with heading ⟨heading⟩. The environment is started with a new numbering sequence. For example, to define a new environment called thm for theorems with a new numbering sequence you would use \newtheorem{thm}{Theorem}.

with an optional argument With the optional argument, the optional argument may be used in different positions. It may be used as the *second* argument and as the *last* argument. The following explains the differences.

- If the optional argument is used as the *second* argument of the \newtheorem command, then you define the environment using \newtheorem{⟨env⟩}[⟨old⟩]{⟨heading⟩}. This defines a new environment ⟨env⟩ with heading text ⟨heading⟩. The environment does not start with a new numbering sequence. Instead, the environment shares its numbering with the existing theorem-style environment ⟨old⟩.
- If the optional argument is used as the *last* argument, you define the environment using \newtheorem{⟨env⟩}{⟨heading⟩}[⟨unit⟩]. This defines a new environment ⟨env⟩ with heading ⟨heading⟩. The argument ⟨unit⟩ should be the name of a sectional unit, for example, chapter, section, ... This defines an environment called ⟨env⟩ with heading ⟨heading⟩ and a new numbering sequence that depends on the sectional unit ⟨unit⟩.

Figure 9.4 provides an example of how the amsthm package may be used to define three theorem-like environments called thm, lem, and def with headings Theorem, Lemma, and Definition. The first two environments are typeset in the style plain. The last environment is typeset in the style definition. The numbering of the environments does not depend on sectional units and is shared.

```
\usepackage{amsmath}
\usepackage{amsthm}


    % Current environment style is plain.
    %% Define environment thm for theorems.
    \newtheorem{thm}{Theorem}
    %% Define environment lemma for lemmas.
    %% Share numbering with thm environment.
    \newtheorem{lemma}[thm]{Lemma}


    % Set environment style to definition.
    \theoremstyle{definition}
    %% Define environment def for definitions.
    %% Share numbering with thm environment.
    \newtheorem{def}[thm]{Definition}
```

Figure 9.4

Using the amsthm package

9.8.4 *Defining Theorem-like Styles*

The command \newtheoremstyle is for defining new amsmath theorem-like environment styles. This command gives you ultimate control over fine typesetting of the environments. Usually the predefined styles plain, definition, and remark suffice. Exact information about the command \newtheoremstyle may be found in the amsthm documentation [American Mathematical Society 2002].

9.8.5 *Proofs*

Writing proofs is done with the proof environment. The environment takes an optional argument for a title of the proof. The environment makes sure that it completes the proof by putting a square ($\square$) at the end of the proof. Adding the square makes it easy to recognise the end of the proof, which is especially nice if proofs get long. Unfortunately, the automatic mechanism for putting the square at the end of the proof doesn't work well if the proof ends in a displayed formula because the square is placed below the last line in the display, which looks dreadful.[1] The following shows how this looks. Never, ever try to do this at home.

Petkovšek, Wilf, and Zeilberger. ...

$$A = B.$$

$\square$

To overcome the problem with the automatic end-of-proof symbol placement, there is also a command called \qedhere. This command may be used to put the square at the end of the last displayed formula.

---

[1]Mind you, every now and then you will see a proof ending like this.

Table 9.3

Math mode dot-like symbols

. `\ldotp`	... `\ldots`	· `\cdotp`
··· `\cdots`	: `\colon`	⋮ `\vdots`
⋱ `\ddots`		

You use the command inside the environment that produces the display.

```
\begin{proof}[Challenge]
The following proves that
$5^{2} = 3^{2} + 4^{2}$:
\[ 5^2 = 25 = 9 + 16
      = 3^2 + 4^2\,.
   \qedhere \]
\end{proof}
```

Challenge. The following proves that $3^2 + 4^2 = 5^2$:

$$5^2 = 25 = 9 + 16 = 3^2 + 4^2. \qquad \square$$

9.9 Mathematical Punctuation

LaTeX provides several commands for typesetting dot-like symbols. Table 9.3 lists LaTeX's built-in commands. Unfortunately, it is not quite clear how these commands should be used. The following provides some guidelines about how these symbols should be used.

`\ldotp`

Used for the definition of `\ldots` [Knuth 1990, page 438]. ☑

`\ldots`

Low dots. Used between commas, and when things are juxtaposed with no signs between them at all [Knuth 1990, page 172]. For example, `$f(x_{1}, \ldots, x_{n} )$` gives you $f(x_1, \ldots, x_n)$ and `$n(n-1)\ldots(1)$` gives you $n(n-1)\ldots(1)$. ☑

`\cdotp`

Used for the definition of `\cdots` [Knuth 1990, page 438]. ☑

`\cdots`

Centred dots. Used between + and – and × signs, between = signs and other binary relational operator signs [Knuth 1990, page 172]. For example, `$x_{1}+\cdots+x_{n}$` gives you $x_1 + \cdots + x_n$. ☑

`\colon`

A punctuation mark [Knuth 1990, page 134]: `$f \colon A \to B$`. ☑

`\ddots`

Diagonal dots. Used in arrays and matrices. ☑

`\vdots`

Vertical dots. Used in arrays and matrices. ☑

Notice that the command `\cdot` also produces a dot. However, this is not used for punctuation. It is generally used in expressions like $(x_1, \ldots, x_n) \cdot (y_1, \ldots, y_n)$ or $f(\cdot)$.

Many symbols in mathematical formulae require a different spacing in a different contexts. The commands that reproduce these symbols are context-unaware and this may result in the wrong spacing. The amsmath package provides several commands to overcome this

```
\ldots Then we have series
    $A_1, A_2, \dotsc$,
regional sum
    $A_1 + A_2 + \dotsb$,
orthogonal product
    $A_1 A_2 \dotsm$,
and infinite integral
    \[  \int_{A_1}
       \!\int_{A_2}
         \dotsi\,.\]
```

...Then we have series $A_1, A_2, \ldots$, regional sum $A_1 + A_2 + \cdots$, orthogonal product $A_1 A_2 \cdots$, and infinite integral

$$\int_{A_1}\!\int_{A_2} \cdots \,.$$

Figure 9.5
Using the mathematical punctuation commands

problem. The following commands are for typesetting dots and sequences of dots.

\dotsc

For dots in combination with commas. ☑

\dotsb

For dots in combination with binary operators/relations. ☑

\dotsm

For multiplication dots. ☑

\dotsi

For dots with integrals. ☑

\dotso

For other dots. ☑

Figure 9.5 demonstrates the effect of these commands. This figure is based on the amsmath documentation [AMS 2002].

9.10 Spacing and Linebreaks

This section provides some information and guidelines related to spacing and linebreaking in math mode. The majority of this section is based on [Knuth 1990, Chapter 18].

9.10.1 *Line Breaks*

LaTeX may break lines after commas in text mode but it doesn't after commas in math mode. This makes sense since you don't want to see a break after the comma in $f(a, b)$. Make sure you keep the commas that are part of formulae inside the dollar expressions in ordinary math mode. The remaining commas should be kept outside. The following example uses these rules correctly.

```
for $x = f( a, b )$, $f( b, c )$,
  or~$f( b, c )$.
```
LaTeX Usage

However, the following is not correct because the second comma is at the sentence level; it is not part of any mathematical expression.

```
for $x = f( a, b ), f( b, c )$,
  or $f( b, c )$.
```
Don't do this at Home

As a final example, the following is also not correct because the first comma is at the sentence level.

```
Let $x, y$, and $z$ be real numbers.                    Don't do this at Home
```

The following corrects the mistake in the previous input.

```
Let $x$, $y$, and $z$ be real numbers.                    LaTeX Usage
```

In display math mode the TeXpert is ultimately responsible for linebreaks and inserting whitespace. This is especially true in environments with alignment positions. The following are a few guidelines.

○ Always insert a thin space (\,) before punctuation symbols at the end of the lines. The reason for doing this is to make the punctuation mark stand out a bit more. If you don't it may get lost in the display.
○ In sums or differences linebreaks should be inserted *before* the plus or minus operator. On the next line you should insert a *qquad* after the alignment position. Here a qquad is equivalent to two quads. One quad has the width of one em, which is equal to the type size. If the continuation line is short you may even consider inserting several qquads. You insert a single quad with the command \quad. A single qquad is inserted with the command \qquad.

```
\begin{align*}                                          LaTeX Usage
  f( x ) & = a + b + c + d \\
         & \qquad + e + f + g\,.
\end{align*}
```

○ Linebreaks in products should occur *after* the multiplication operator.

```
\begin{align*}                                          LaTeX Usage
  f( x ) & = a \times b \times c \times d \times\\
         & \qquad e \times f \times g\,.
\end{align*}
```

Clearly, breaking equations by hand is prone to errors. The breqn package [Høgholm 2008] automates line breaking.

9.10.2 Conditions

In ordinary math mode, you should put an extra space for conditions following equations. This makes the conditions stand out a bit more.

```
The Fibonacci numbers satisfy                           LaTeX Usage
  $F_{n} = F_{n - 1} + F_{n - 2}$, \ $n \geq 2$.
```

However, it is probably better to turn the previous example into a proper sentence as follows.

```
The Fibonacci numbers satisfy                           LaTeX Usage
  $F_{n} = F_{n - 1} + F_{n - 2}$, for~$n \geq 2$.
```

If you must add a condition to a formula in display math mode then the two should be separated with a single qquad.

```
\[ z^{m} G( z ) = \sum_{n} g_{n - m} z^{n}\,,
    \qquad\text{integer $m \geq 0$}\,. \]
```
LaTeX Usage

You may also put the condition in parentheses. However, if you do this, you omit the comma before the condition.

```
\[ z^{m} G( z ) = \sum_{n} g_{n - m} z^{n}
    \qquad\text{(integer $m \geq 0$)}\,. \]
```
LaTeX Usage

9.10.3 *Physical Units*

Physical units should be typeset in roman. In expressions of the form ⟨number⟩ ⟨unit⟩, you insert a thin space between the number and the unit: ⟨number⟩\,⟨unit⟩. The following is a concrete example.

```
$g = 9.8\,\mathrm{m}/\mathrm{s}^{2}$
```
LaTeX Usage

The siunitx package provides support for typesetting units. Using the package you write \SI{9.8}{\metre\per\second\squared}. This gives you $9.8\,\mathrm{m\,s^{-2}}$ as standard, or $9.8\,\mathrm{m/s^2}$ by setting per=slash with the \sisetup macro. More information about the siunitx package may be found in the package documentation [Wright 2011].

9.10.4 *Sets*

Sets come in two flavours. On the one hand there are "ordinary" sets, the definitions of which do not depend on conditions: $\{1\}$, $\{3, 5, 6\}$, and so on. On the other hand there are "guarded sets" whose definitions do depend on conditions: $\{2n : n \in \mathbb{N}\}$ and so on.

ordinary sets For ordinary sets there is no need to add additional spacing after the opening brace and before the closing brace.

```
The natural numbers, $\mathbb{N}$, are defined
$\mathbb{N} = \{ 0, 1, 2, \ldots \}$.
```
LaTeX Usage

guarded sets For guarded sets you insert a thin space after the opening and before the closing brace. The use of a thin space before and after the colon is not recommended by Knuth [1990], but it may be argued that it makes the result easier to read.

```
The even numbers, $E$, are defined
$E = \left\{\, 2 n \,:\, n \in \mathbb{N} \,\right\}$.
```
LaTeX Usage

If you don't like the colon then you should write the following.

```
The even numbers, $E$, are defined
$E = \left\{\, 2 n \,\mid\, n \in \mathbb{N}
\,\right\}$.
```
LaTeX Usage

Table 9.4
This table demonstrates the effect of spacing commands. The first two columns list positive spacing commands, the next two columns demonstrate the effect of the \hphantom command, and the last two columns list negative spacing commands. The effect of the command is the distance from the right arrow tip to the left arrow tip. The distance is negative if the arrow tips overlap horizontally.

Positive Spacing		\hphantom		Negative Spacing	
\,		\hphantom{M}		\!	
\thinspace		M		\negthinspace	
\:		\hphantom{z^n}		\negmedspace	
\medspace		z^n		\negthickspace	
\;					
\thickspace					
\quad					
\qquad					

9.10.5 *More Spacing Commands*

Table 9.4 demonstrates the effect of the horizontal spacing commands. The command \hphantom in this table is related to the command \phantom. It results in a box that has zero height and a width that is equal to the width of the argument of the command.

Chapter 10
Algorithms and Listings

ALGORITHMS ARE UBIQUITOUS in computer science papers. Knowing how to present your algorithms increases the chance of getting your ideas across. This chapter explains how to typeset pseudo-code with the `algorithm2e` package and how to present verbatim program listings with the `listings` package.

10.1 Presenting Pseudo-Code with `algorithm2e`

This section introduces `algorithm2e`, which is a popular package for typesetting algorithms in pseudo-code. The remainder of this section explains some important package aspects. The explanation is based on the package documentation [Fiorio 2004]. Notice that if you don't like the `algorithm2e` package, you can always fall back to the `tabbing` environment, which is explained in Section 2.19.6.

10.1.1 *Loading `algorithm2e`*

Loading `algorithm2e` properly may save time when writing your algorithms. An important option is `algo2e`. This option renames the environment `algorithm` to `algorithm2e` so as to avoid name clashes with other packages.

```
\usepackage[algo2e]{algorithm2e}
```
L^AT_EX Usage

There are several options that affect algorithm appearance. The following three options control block typesetting.

noline This option results in blocks that are typeset without vertical lines marking the scope of the blocks. The picture on the left of Figure 10.1 demonstrates the effect of this option for a simple conditional statement.

lined This option draws a vertical line indicating the scope of the block. The keyword that indicates the end of the block is still typeset. The picture in the centre of Figure 10.1 demonstrates the effect of this option for a simple conditional statement.

vlined This option also results in vertical lines indicating the scopes of each block. However, this time the end of each scope is indicated by a little "bend" in the line. With this option the keyword indicating the end of the block is not typeset. The picture on the right of Figure 10.1 demon-

Figure 10.1

Algorithm style. The options noline (left), lined (centre), and vlined (right) control the output of algorithm2e, with vlined saving the most vertical space.

if ⟨cond⟩ **then**
    ⟨stuff⟩
end
	if ⟨cond⟩ **then**
	  | ⟨stuff⟩
	end
		if ⟨cond⟩ **then**
		  └ ⟨stuff⟩

strates the effect of this option for a simple conditional statement. Compared to the other options, this option is more economical in terms of saving vertical space. When writing a paper this may make the difference between making the page count and overrunning it.

The algorithm2e package has many more options. For further information, read the package documentation. Examples in the remainder of this section are typeset with the option vlined.

10.1.2 *Basic Environments*

The algorithm2e package defines a number of basic environments. Each of them is typeset in a floating environment like a figure or table. The \caption option is available in the body of the environment and works as explained in Section 6.6. The command \listofalgorithms may be used to output a list of the algorithms with a caption. This is usually done in the document preamble. The package option dotocloa adds an entry for the list of algorithms in the table of contents. The following are the environments:

algorithm Typesets its body as an algorithm.

algorithm* Typesets its body as an algorithm in a two-column document. The resulting output occupies two columns.

procedure Typesets its body as a procedure. Compared to algorithm there are a couple of differences:
  ○ The caption starts by listing 'Procedure ⟨name⟩.'
  ○ The caption must start with ⟨name⟩(⟨arguments⟩).

procedure* Typesets its body as a procedure in a two-column document. This environment works like procedure except that the resulting output occupies two columns.

function Typesets its body as a function. This environment works like procedure.

function* Typesets its body as a function in a two-column document. This environment works like function but the resulting output occupies two columns.

Each environment may be positioned using its optional argument, which may be any combination of p, t, b, or h, and each has the usual meaning. This positioning mechanism is explained in Section 6.6. The option H is also allowed and means "definitely here." The best choice is to omit the optional argument or use tbp:

```
\begin{algorithm2e}[H]
\KwIn{
Integers $a \geq 0$ and $b \geq 0$}

\KwOut{\textsc{Gcd}
of $a$ and $b$}
\While{$b \neq 0$}{
    $r \leftarrow a \bmod b$\;
    $a \leftarrow b$\;
    $b \leftarrow r$\;
}
\caption{Euclidean Algorithm}
\end{algorithm2e}
```

Input: Integers
$a \geq 0$ and
$b \geq 0$
Output: GCD of a
and b
while $b \neq 0$ **do**
$\quad | \quad r \leftarrow a \bmod b;$
$\quad | \quad a \leftarrow b;$
$\quad | \quad b \leftarrow r;$

Algorithm 1: Euclidean Algorithm

Figure 10.2
Using algorithm2e

```
\begin{algorithm2e}[tbp]
…
\end{algorithm2e}
```

LaTeX Usage

Figure 10.2 demonstrates some of the algorithm2e functionality. Notice that the semicolons are typeset with the command \;.

10.1.3 Describing Input and Output

The algorithm2e package has several commands for describing algorithm input and output. It also provides a mechanism to add keywords and define a style for classes of keywords. This section briefly explains the main commands for describing algorithm input and output.

\KwIn{⟨input⟩}
This typesets the Input label followed by ⟨input⟩. Figure 10.2 demonstrates how this works.   ☑

\KwOut{⟨output⟩}
This typesets the Out label followed by ⟨output⟩.   ☑

\KwData{⟨data⟩}
This typesets the Data label followed by ⟨data⟩.   ☑

\KwResult{⟨output⟩}
This typesets the Result label followed by ⟨output⟩.   ☑

\KwRet{⟨return value⟩}
This typesets the Ret label followed by ⟨return value⟩.   ☑
   It is possible to redefine the values of the default labels in the previous list. For example, the command \SetKwInput{KwIn}{⟨label⟩} redefines the value of the Input label to ⟨label⟩. Fiorio [2004, Section 9.8] explains how to redefine the other labels.

10.1.4 Conditional Statements

The algorithm2e package defines a large array of commands for typesetting conditional statements. This includes commands for typesetting one-line statements. The remainder of this section explains some of the commands for typesetting simple multi-line conditional statements. Information about other conditional commands may be found in the the package documentation.

\If(⟨comment⟩){⟨condition⟩}{⟨clause⟩}

This typesets a single conditional statement with condition ⟨condition⟩ and final then clause ⟨clause⟩. The argument in parentheses describes a comment. This argument is optional and may be omitted. The following is an example of the resulting output. The comment has been omitted.

```
\If{⟨condition⟩}
    {⟨clause⟩}
```
if ⟨condition⟩ **then**
└ ⟨clause⟩   ☑

\uIf(⟨comment⟩){⟨condition⟩}{⟨clause⟩}

This works as \If only this time it is assumed that ⟨clause⟩ is not the final clause. The following is the resulting output.

```
\uIf{⟨condition⟩}
    {⟨clause⟩}
```
if ⟨condition⟩ **then**
│ ⟨clause⟩   ☑

\ElseIf(⟨comment⟩){⟨condition⟩}{⟨clause⟩}}

This typesets a conditional else clause with condition ⟨condition⟩ and final else if clause ⟨clause⟩.

```
\ElseIf{⟨condition⟩}
      {⟨clause⟩}
```
else if ⟨condition⟩ **then**
└ ⟨clause⟩   ☑

\uElseIf(⟨comment⟩){⟨condition⟩}{⟨clause⟩}}

This typesets a conditional else clause with condition ⟨condition⟩ and non-final else clause ⟨clause⟩.

```
\eUlseIf{⟨condition⟩}
        {⟨clause⟩}
```
else if ⟨condition⟩ **then**
│ ⟨clause⟩   ☑

\eIf(⟨comment⟩){⟨condition⟩}{⟨then clause⟩}(⟨comment⟩){⟨else clause⟩}}

This typesets the if else clause with condition ⟨condition⟩ with then clause ⟨then clause⟩ and final else clause ⟨else clause⟩. As suggested by the notation, both ⟨comment⟩ arguments are optional.

```
\eIf{⟨condition⟩}
    {⟨then clause⟩}
    {⟨else clause⟩}
```
if ⟨condition⟩ **then**
│ ⟨then clause⟩
else
└ ⟨else clause⟩   ☑

```
\begin{algorithm2e}[tbp]
\uIf{$a < 0$}{
    \tcp{$a < 0$}
} \uElseIf{$a = 0$}{
    \tcp{$a = 0$}
} \lElse\eIf{$a = 1$}{
    \tcp{$a = 1$}
} {
    \tcp{$a > 1$}
}
\end{algorithm2e}
```

if $a < 0$ **then**
  | // $a < 0$
else if $a = 0$ **then**
  | // $a = 0$
else if $a = 1$ **then**
  | // $a = 1$
else
  | // $a > 1$

Figure 10.3
Typesetting conditional statements

\lElse

This typesets the word else. This is mainly useful in combination with \eIf. ☑

Figure 10.3 typesets a complex if statement. The command \tcp in Figure 10.3 typesets its argument as a C++ comment.

10.1.5 *The Switch Statement*

This section briefly explains algorithm2e commands for typesetting switch statements.

\Switch(⟨comment⟩){⟨value⟩}{⟨cases⟩}

This typesets the first line and the braces for the body of the switch statement. The following is the resulting output.

```
\Switch{⟨value⟩}
    {⟨cases⟩}
```

switch ⟨*value*⟩ **do**
  | ⟨cases⟩ ☑

\Case(⟨comment⟩){⟨condition⟩}{⟨statements⟩}

This typesets the final case of the switch statement. The following is the resulting output.

```
\Case{⟨condition⟩}
    {⟨statements⟩}
```

case ⟨*condition*⟩
  | ⟨statements⟩ ☑

\uCase(⟨comment⟩){⟨condition⟩}{⟨statements⟩}

This also typesets a case of the switch statement, but here it is assumed the case is not the last case of the switch statement. The following is the resulting output.

```
\uCase{⟨condition⟩}
    {⟨statements⟩}
```

case ⟨*condition*⟩
  | ⟨statements⟩ ☑

Figure 10.4

Using algorithm2e's switch statements

```
\begin{algorithm2e}[tbp]
\Switch{order}{
    \uCase{bloody mary}{
        Add tomato juice\;
        Add vodka\;
        break\;
    }
    \uCase{hot whiskey}{
        Add whiskey\;
        Add hot water\;
        Add lemon and cloves\;
        Add sugar or honey to taste\;
        break\;
    }
    \Other{Serve water\;}
}
\end{algorithm2e}
```

switch *order* **do**
   case *bloody mary*
      | Add tomato juice;
      | Add vodka;
      | break;
   case *hot whiskey*
      | Add whiskey;
      | Add hot water;
      | Add lemon and cloves;
      | Add sugar or honey to taste;
      | break;
   otherwise
      | Serve water;

\Other(⟨comment⟩){⟨statements⟩}

This typesets the default case of the switch statement. The following is the resulting output.

```
\Other{⟨statements⟩}
```

otherwise
| ⟨statements⟩   ☑

The example in Figure 10.4 shows how to typeset a complete switch statement. The example also shows how to make a drink or two.

10.1.6 *Iterative Statements*

The algorithm2e package typesets several iterative statements, including while, for, foreach-based, and repeat-until statements. The following explains these commands.

\For(⟨comment⟩){⟨condition⟩}{⟨body⟩}

This typesets a basic for statement with a condition ⟨condition⟩ and body ⟨body⟩.

```
\For{⟨condition⟩}
    {⟨body⟩}
```

for ⟨condition⟩ **do**
| ⟨body⟩   ☑

\ForEach(⟨comment⟩){⟨condition⟩}{⟨body⟩}

This typesets a foreach statement with a condition ⟨condition⟩ and body ⟨body⟩.

```
\ForEach{⟨condition⟩}          foreach ⟨condition⟩ do
        {⟨body⟩}                 └ ⟨body⟩                    ☑
```

\While(⟨comment⟩){⟨condition⟩}{⟨body⟩}

This typesets a while statement with condition ⟨condition⟩ and body ⟨body⟩.

```
\While{⟨condition⟩}            while ⟨condition⟩ do
      {⟨body⟩}                   └ ⟨body⟩                    ☑
```

\Repeat(⟨comment⟩){⟨condition⟩}{⟨body⟩}(⟨comment⟩)

This typesets a repeat-until statement with and body ⟨body⟩. The first comment is put on the repeat line. The second comment is put on the until line. The following is a short example.

```
\Repeat{⟨condition⟩}           repeat
       {⟨body⟩}                  | ⟨body⟩                     ☑
                               until ⟨condition⟩;
```

10.1.7 *Comments*

This Section, which concludes the discussion of the algorithm2e package, explains how to typeset comments. Comments are defined in a C and C++ style. For a given language there are different styles of comments. The command for typesetting C comments is \tcc, for typesetting C++ comments use tcp. The following explains the tcp command. The \tcc command works the same.

\tcp{⟨comment⟩}

This typesets the comment ⟨comment⟩, which may consist of several lines. Comment lines should be delimited with newlines (\\).

```
\tcp{⟨line one⟩}\\            // ⟨line one⟩
    {⟨line two⟩}              // ⟨line two⟩                  ☑
```

\tcp*{⟨comment⟩}

This typesets a side comment ⟨comment⟩ right justified. The command \tcp*[r]{⟨comment⟩} works analogously.

```
⟨statement⟩
   \tcp*{⟨comment⟩}           ⟨statement⟩; // ⟨comment⟩       ☑
```

\tcp*[l]{⟨comment⟩}

This typesets a side comment ⟨comment⟩ left justified.

```
⟨statement⟩
  \tcp*[l]{⟨comment⟩}          ⟨statement⟩;// ⟨comment⟩      ☑
```

\tcp*[h]{⟨comment⟩}

This typesets the comment ⟨comment⟩ left justified in place (here).

```
\If(\tcp*[h]{⟨comment⟩})
   {⟨condition⟩}              if ⟨condition⟩ then // ⟨comment⟩
   {⟨statement⟩}              └ ⟨statement⟩                  ☑
```

\tcp*[f]{⟨comment⟩}

This typesets the comment ⟨comment⟩ right justified in place (here).

```
\If(\tcp*[f]{⟨comment⟩})
   {⟨condition⟩}              if ⟨condition⟩ then // ⟨comment⟩
   {⟨statement⟩}              └ ⟨statement⟩                  ☑
```

10.2 The listings Package

The listings package is one of the nicer packages for creating format-ted output. The remainder of this section is a brief example-driven introduction to the package. More information may be found in the package documentation [Heinz, and Moses 2007].

The listings package supports the typesetting of verbatim listings. The package provides support for several languages, including ANSI C, and ANSI C++, Eiffel, HTML, Java, PHP, Python, LaTeX, and XML. The package supports user-defined styles for keywords and identifiers. Two methods are provided for specifying a listing.

environment An environment called lstlisting for specifying a listing in the body of the environment.

command A command called \lstinputlisting for creating a listing from a source file. The required argument of this command is the name of the source code file. The optional argument determines the options.

Both the environment and the command take an optional argument in the form of a ⟨key⟩=⟨value⟩ list, for overriding the default settings. The package also provides a command for setting new defaults. The resulting algorithm may be typeset at the current position or as a float-ing algorithm with a number and caption. The package also provides a command called \listoflistings for typesetting a list of numbered listings.

Figure 10.5 shows the lstlistings environment. The resulting output is presented in Figure 10.6. Note that not all of the body of the environment is typeset and that the resulting numbers are generated automatically. The following explains the relevant options.

```
\begin{lstlisting}[language=Java
                  ,gobble=3
                  ,numbers=left
                  ,firstline=2
                  ,lastline=4
                  ,firstnumber=2
                  ,caption=Hello World.
                  ,label=example]
public class Greetings {
␣␣␣public static void main( String[] args ) {
␣␣␣    System.out.println( "Hello world!" );
␣␣␣}
}
\end{lstlisting}
```

Figure 10.5
Creating a partial listing with
the listings package

```
2 ␣public static void main( String[] args ) {
3 ␣    System.out.println( "Hello world!" );
4 ␣}
```
Listing 1. Hello world.

Figure 10.6
Listing created from input in
Figure 10.5

language This specifies the programming language. Possible values are C, [ANSI]C, C++, [ANSI]C++, HTML, Eiffel, HTML, Java, PHP, Python, LaTeX, and XML.

gobble This determines the number of characters that should be removed from the start of the input lines. The default value is 0.

numbers This is used to control the placement of numbers. Possible values are none (default) for no numbers, left for numbers on the left, and right for numbers on the right.

firstline The value of this option determines the number of the first input line that is typeset. It may be useful to skip a number of lines at the start of the source. The default value is 1.

lastline This option determines the number of the last input line that is typeset. The default value is the number of lines in the input.

firstnumber This is the first line number in the output.

caption This determines the caption of the typeset listing.

label This determines the label reference the listing with the \ref command.

As already stated, the listings package provides a command for specifying new default option values. The name of this command is \lstset and its required argument is a list of ⟨key⟩=⟨value⟩ arguments specifying the new default values for the options.

Figure 10.7 provides an example that overrides some of the default settings. Some of these options have been explained before. The remaining options work as follows.

keywordstyle The value of this option is a series of declarations that determine how keywords are typeset. The declarations \bfseries\ttfamily in Fig-

Figure 10.7

Setting new defaults with the
\lstset command

```
\lstset{language=Java%
       ,keywordstyle=\bfseries\ttfamily%
       ,stringstyle=\ttfamily%
       ,identifierstyle=\ttfamily\itshape%
       ,showspaces=false%
       ,showstringspaces=true%
       ,numbers=left%
       ,float%
       ,floatplacement=tbp%
       ,captionpos=b}
```

ure 10.7 result in bold face keywords that are typeset in a monospaced font.

stringstyle The value of this option is a series of declarations that determine how characters are typeset in strings. The declaration \ttfamily in Figure 10.7 result in string characters that are typeset in a monospaced font.

identifierstyle The value of this option is a series of declarations that determines how identifiers are typeset. The declarations \ttfamily\itshape in Figure 10.7 result in identifiers that are typeset in a monospaced italic font.

showspaces If the value of this option is true then spaces are typeset as visual spaces. The default value is false.

showstringspaces If the value of this option is true then spaces in strings are typeset as visual spaces. The default value is false.

float If this option is provided then the listing is typeset as a float.

floatplacement The value of this option determines the float placement. It can be any sequence of characters in tbph.

captionpos This determines the position of the caption. Possible values are t (top) and b (bottom).

PART V

Automation

Oil paint and charcoal on canvas (17/06/09), 152 × 213 cm
Work included courtesy of Billy Foley
© Billy Foley (www.billyfoley.com)

CHAPTER 11
Commands and Environments

THIS CHAPTER STUDIES user-defined commands and environments in LaTeX2_ε, which is the LaTeX implementation that was current at the time this chapter was written.

11.1 Some Terminology

This section briefly introduces some terminology for the remainder of this chapter. Throughout this chapter we shall use the word parameter for an argument or parameter of a macro or command. An *actual parameter* is a parameter that is passed to an existing command. A *formal parameter* is a placeholder in the definition of a command for an actual parameter.

For example, consider the mathematical function definition that is given by

$$f : \mathbb{N} \to \mathbb{N} \tag{11.1}$$
$$x \mapsto 2x. \tag{11.2}$$

The definition of $f(\cdot)$ has two parts. Equation 11.1 is the first part of the definition; it determines the signature of the function. In LaTeX there is no equivalent for the signature. Equation 11.2 is the second part of the definition; it defines the semantics of the function. The semantics may be regarded as an input-output transformation, with x determining the input parameter and $2x$ determining the output. The input parameter x defines a name that may be used in the output expression. This name acts as the first (and only) formal parameter in the definition of $f(\cdot)$. Using the function, one may write $f(1)$, or even $f(x)$, assuming that x has a proper context. Here 1 and x are actual parameters. The result of the expression $f(a)$, where a is an actual parameter, can be found by substituting the actual parameter a for the formal parameter x in the method's definition: $f(a) = (2x)|_{x=a}$, which gives us $2a$. Effectively, formal and actual parameters in LaTeX work in a similar way.

11.2 Advantages and Disadvantages

LaTeX is a programmable typesetting engine. Commands are the key

to controlling your document. The advantages of using commands in
LaTeX are similar to the advantages of using functions and procedures
in high-level programming languages. However, LaTeX commands
also have disadvantages. We shall first study advantages and then
disadvantages. The following are some advantages.

software engineering Tedious tasks can be automated. This has the following advantages.

reusability Commands that are defined once can be reused several times.

simplicity Carrying out a complex task using a simple command with a well-understood interface is much easier and leads to fewer errors.

refinement You can stepwise refine the implementation of certain tasks. This lets you postpone certain decisions. For example, if you haven't been able to decide how to typeset certain symbols that serve a certain purpose, then you may start typesetting them using a command that typesets them in a simple manner. This lets you start writing the document in terms of high-level notions (procedural markup). By refining the command at a later stage, you can fine-tune the typesetting of all the relevant symbols.

maintainability This advantage is related to the previous item. Unforeseen changes in requirements can be implemented easily by making a few local changes.

consistency Typesetting entities using carefully chosen commands guarantees a consistent appearance of your document. For example, if you typeset your pseudo-code identifiers using a pseudo-code identifier typesetting command in a "pseudo-code identifier" style, then your identifiers will have a consistent feel.

computing Tasks and results may be computed depending on document options. This has the following advantages.

style control Things may be typeset in a style that depends on class or package options. For example, the article class typesets the main text in 10 pt by default but providing a 12pt option gives you a 12 pt size.

content control Commands may result in different output depending on a global mode. For example, consider the beamer class, which lets you prepare a computer presentation and lecture notes in the same input. In presentation mode the beamer class results in a computer presentation but in article mode it may result in lecture notes. You can share text for the notes and the presentations but you can also hide text in the notes or in the presentation. This is very a strong feature because it allows sharing and guarantees consistency between the notes and the presentation.

typeset results This issue is related to the previous item. LaTeX can do basic arithmetic, can branch and iterate, and can typeset the *results* of computations. For example, the lipsum package

provides a command \lipsum[⟨number₁⟩-⟨number₂⟩] that typesets the *Lorem ipsum* paragraphs ⟨number₁⟩ – ⟨number₂⟩. You can
easily extend this command to make it repeat the paragraphs
a given number of times. As another example, again consider
the beamer class, which can uncover items in an itemized list
one at a time. Uncovering the items results in several partial
and one final page from the single frame. As a final example,
the calctab package provides the spreadsheet functionality
with computation rules for output cells in tables.

The following are some disadvantages of LATEX commands.

namespace limitation TEX allows local definitions at the group level but its namespace is flat at the top level. All top-level commands are global. This
is arguably the greatest problem. With thousands of packages and
classes this requires that package and class implementors have to be
careful to avoid name clashes.

parameter limitation There are two problems related to the parameters of the
commands.

number TEX sadly does not allow more than nine parameters per
macro. It may be argued that commands that require more
than nine parameters are not well-designed, but this does not
make the restriction less arbitrary.

names This is probably the source of the previous disadvantage.
When Knuth implemented TEX, he decided to use numbers
as formal macro parameters. The first is called #1, the second is called #2, and so on. Needless to say that this makes it
extremely easy for TEX to parse and recognise formal parameters, but this takes away the possibility to choose meaningful
names for the formal parameters, makes it difficult to understand the implementation of the commands, and makes it
easy to make mistakes.

11.3 User-defined Commands

This section studies command definitions. Section 11.3.1 explains
how to define and redefine commands that take no parameters. Section 11.3.2 explains how to define and redefine commands that do take
parameters, and Section 11.3.3 explains the difference between *fragile*
and *robust* commands. Section 11.3.4 explains how to define robust
commands and make existing commands robust.

11.3.1 *Defining Commands Without Parameters*

LATEX has several commands to define new commands. The following
commands define and redefine commands that take no parameters.

\newcommand⟨cmd⟩{⟨subst⟩}

This defines a new command, ⟨cmd⟩, with substitution text ⟨subst⟩. In
TEX parlance ⟨cmd⟩ is called a *control sequence*. A LATEX control sequence

starts with a backslash and is followed by a non-empty sequence of characters—usually letters. The new command does not take any parameters. The substitution text ⟨subst⟩ is substituted for each occurrence of ⟨cmd⟩ that is expanded by the Expansion Processor. This does not include all occurrences. For example ⟨cmd⟩ is not expanded if it occurs in the substitution text of other LaTeX definitions at definition time. Section 11.4 provides a more detailed description of the expansion of LaTeX commands.　　　　　　　　　　　　　　　☑

`\renewcommand⟨cmd⟩{⟨subst⟩}`

This redefines the command ⟨cmd⟩, which should be an existing command. The resulting command has substitution text ⟨subst⟩ and does not take any parameters.　　　　　　　　　　　　　　　☑

The following example defines a user-defined command \CTAN and uses it in the body of the document environment.

```
\documentclass{article}                                    LaTeX Usage
\newcommand\CTAN{Comprehensive \TeX{} Archive Network}
\begin{document}
  I always download my packages from the \CTAN.
  The \CTAN{} is the place to be.
\end{document}
```

The substitution text of the command is 'Comprehensive \TeX{} Archive Network.' Given this definition LaTeX substitutes the substitution text 'Comprehensive \TeX{} Archive Network' for \CTAN each time \CTAN is used. The following is the resulting output.

> I always download my packages from the Comprehensive TeX Archive Network. The Comprehensive TeX Archive Network is the place to be.　　LaTeX Output

Finally, there are also `\newcommand*` and `\renewcommand*` commands. They work just as `\newcommand` and `\renewcommand` but their parameters may not contain paragraph tokens.

11.3.2 Defining Commands With Parameters

Defining commands with parameters is almost the same as defining commands without parameters. However, this time you have to provide the substitution text in terms of the formal parameters. The following are the related commands.

`\newcommand⟨cmd⟩[⟨digit⟩]{⟨subst⟩}`

As before, this defines a new command, ⟨cmd⟩, with substitution text ⟨subst⟩. This time the command takes ⟨digit⟩ parameters. The number of parameters should be in the range 1–9. The i-th formal parameter is referred to as #i in the substitution text ⟨subst⟩. When substituting ⟨subst⟩ for ⟨cmd⟩ TeX's Expansion Processor also substitutes the i-th actual parameter for #i in ⟨subst⟩, for $1 \leq i \leq$ ⟨digit⟩. It is not allowed to use #i in ⟨subst⟩ if i < 1 or ⟨digit⟩ < i.　　☑

```
\usepackage{multind}
\makeindex{command}
\makeindex{package}

\newcommand\MonoIdx[2][command]{
    \texttt{#2}%
    \index{#1}{\texttt{#2}}%
}

\begin{document}
  ...The command
    \MonoIdx{\textbackslash_MakeRobustCommand}
    is provided by the package
    \MonoIdx[package]{makerobust}. ...
    \printindex{command}{Index of Commands}
    \printindex{package}{Index of Packages}
\end{document}
```

Figure 11.1

User-defined commands

\renewcommand⟨cmd⟩[⟨digit⟩]{⟨subst⟩}

This redefines ⟨cmd⟩ as a command with ⟨digit⟩ parameters and substitution text ⟨subst⟩.   ☑

   The standard way to define a command with an optional parameter
is as follows. By default the first parameter is optional.

\newcommand⟨cmd⟩[⟨digit⟩][⟨default⟩]{⟨subst⟩}

This defines a new control sequence, ⟨cmd⟩, with substitution text
⟨subst⟩. As before the command takes ⟨digit⟩ parameters. However,
this time the first parameter (#1) is optional. If present it should be
enclosed in square brackets. If the optional parameter is omitted then
it is assigned the value ⟨default⟩.   ☑

\renewcommand⟨cmd⟩[⟨digit⟩][⟨default⟩]{⟨subst⟩}

This redefines the existing command ⟨cmd⟩. Essentially, it is equivalent
to \newcommand except that ⟨cmd⟩ must be an existing command. The
value of ⟨digit⟩ may differ from the number of parameters of the
existing command.   ☑

   The LaTeX program in Figure 11.1 uses multiple index files and
defines a user-defined command \MonoIdx that typesets its second
parameter in monospaced font and writes information about it to
these index files. The optional parameter is used to determine the
name of the index file.

11.3.3 *Fragile and Robust Commands*

Having dealt with advantages and disadvantages of LaTeX commands
and knowing how to define them, we're ready to study *fragile* and *robust*
commands. The reason for studying them is that they are a common
cause of errors, which are caused by command side-effects. To make
things worse these errors may occur in subsequent LaTeX sessions and
at seemingly unrelated locations. These errors are difficult to deal

with—especially for novice users. Some of these issues are related to the notions of *moving arguments* and *fragile* and *robust* commands. The remainder of this section explains how to deal with fragile commands in moving arguments and how to avoid these common errors.

A *moving arguments* of a command is saved by the command to be reread later on. Examples of moving arguments are parameters that appear in the Table of Contents, in the Table of Figures, in indexes, and so on. For example, the \caption command defines captions of tables and figures. It writes these captions to the list of tables (.lot) and the list of figures (.lof) files. LaTeX rereads the list of tables file and list of figures file when it typesets the list of figures and the list of tables.

Moving parameters are expanded before they are saved. Sometimes the expansion leads to invalid TeX being written to a file. When this invalid TeX is reread in a subsequent session this may cause errors.

A command that expands to valid TeX is called *robust*. Otherwise it is called *fragile*.

The command \protect protects commands against expansion. If \protect\command is saved then this saves \command. This allows you to protect fragile commands in moving arguments. In effect this postpones the expansion of \command until it is reread.

11.3.4 *Defining Robust Commands*

The following commands are related to defining robust commands and making existing commands robust.

\DeclareRobustCommand⟨cmd⟩{⟨subst⟩}

This defines ⟨cmd⟩ as a robust command without parameters and substitution text ⟨subst⟩. ☑

\DeclareRobustCommand⟨cmd⟩[⟨digit⟩]{⟨subst⟩}

This defines ⟨cmd⟩ as a robust command with substitution text ⟨subst⟩ and ⟨digit⟩ parameters. ☑

\DeclareRobustCommand⟨cmd⟩[⟨digit⟩][⟨default⟩]{⟨subst⟩}

This defines ⟨cmd⟩ as a robust command with substitution text ⟨subst⟩ and ⟨digit⟩ parameters, one of which is optional with default value ⟨default⟩. ☑

\MakeRobustCommand⟨cmd⟩

This turns the existing command ⟨cmd⟩ into a robust command. The \MakeRobustCommand command is not standard but is provided by the package makerobust. ☑

11.4 Commands and Parameters

This section explains how LaTeX applies commands to parameters. Recall from Chapter 1 that TeX's Input Processor turns the input program into a token sequence. After this there are two kinds of tokens:

character tokens A character token represents a single character in the input.

control sequence tokens A control sequence tokens corresponds to a command. It represents a sequence of characters in the input that starts with a backslash and continues with letters.

Recall from Chapter 1 that TEX's Expansion and Execution Processors rewrite token sequences to token sequences. Both of these processors can distinguish between character and control sequence tokens, which makes it easy to recognise tokens that correspond to commands.

It remains to explain how TEX parses parameters. This is slightly more difficult. There are two kinds of parameters, which we shall refer to as *primitive* and *compound* parameters.

primitive parameters A primitive parameter is a single character or control sequence token. The tokens of the opening and closing brace are not allowed.

compound parameters A compound parameter is a brace-delimited group in the input. The token at the start of the group is that of an opening brace ({) and the token at the end of the group is that of a closing brace (}). Within the sequence brace pairs should be balanced. Most of the time you will use compound parameters. The value of a compound parameter is the sequence of tokens "in" the group, that is, the token sequence without the opening and closing brace tokens [Knuth 1990, pages 204–205]. For example, given a command \single that takes one single parameter, the actual parameter of \single{␣ab{c}} is '␣ab{c}.'

The remainder of this section provides examples of command expansion. We shall start with a simple example that involves primitive parameters only, and continue with a more complex example that involves both primitive and compound parameters.

The following explains what happens with primitive parameters. Let's assume we have two user-defined commands called \swop and \SWOP that are defined as follows.

```
\newcommand\swop[2]{#2#1}
\newcommand\SWOP[2]{#2#1}
```
LATEX Usage

Both commands do the same but, for sake of the example, they have different names. Each command takes two parameters and "outputs" the second actual parameter followed by the first actual parameter. With these definitions, \swop2\SWOP31 gives us 321. To see why this happens, notice that the input starts with the command \swop, which takes two parameters. The next two tokens are 2 and \SWOP. Expanding \swop2\SWOP reverses the order of the actual parameters of \swop and results in the token sequence \SWOP2, which is substituted for \swop2\SWOP in \swop2\SWOP31. After this rewriting step the token sequence is \SWOP231. Expanding this token sequence once more gives us 321, which is completely expanded, cannot be expanded any further, and completes the rewriting process.

The following is a more complex example. Let's assume we have the LATEX program that is listed in Figure 11.2. The program defines

Figure 11.2

A program with user-defined combinators

```
\documentclass{article}

\newcommand\K[2]{#1}
\newcommand\S[3]{#1#3{#2#3}}
\newcommand\I{\S\K\K}
\newcommand\X{\S{\K{\S\I}}{\S{\K\K}\I}}

\begin{document}
  \X abc
\end{document}
```

four commands \K, \S, \I, and \X. The first three commands correspond to the combinators K, S, and I from Moses Schönfinkel and Haskell Curry's combinatory logic. They may be described as follows: $K\langle A\rangle\langle B\rangle \mapsto \langle A\rangle$, $S\langle A\rangle\langle B\rangle\langle C\rangle \mapsto \langle A\rangle\langle C\rangle(\langle B\rangle\langle C\rangle)$, and $I \mapsto S\,K\,K$. If you study the TEX definition of the command \X, you may notice that it does not have any formal parameters. It may therefore come as a surprise that it correspond to a combinator, X, that swops its parameters, i.e., $X\langle A\rangle\langle B\rangle \mapsto \langle B\rangle\langle A\rangle$. Still this makes perfect sense and the remainder of this section explains why.

Knowing that \X is a combinator that swops its parameters we should be able to predict the output of our program—it should be 'bac.' Let's see if we can explain this properly. Table 11.1 illustrates the expansion process. The second column of the table lists the output of the Expansion Processor, the third column lists the current input stream of the Expansion Processor, and the first column lists the number of the reductions. The subscripts of the tokens in the input stream correspond to the nesting level of the groups.

The first reduction is that of \X to its substitution text. It does not involve any parameter. Reduction 2 is a reduction of the form $\backslash S\langle A\rangle\langle B\rangle\langle C\rangle \mapsto \langle A\rangle\langle C\rangle\{\langle B\rangle\langle C\rangle\}$, where $\langle A\rangle$ and $\langle B\rangle$ are the top-level groups in the input and $\langle C\rangle$ is the character token of the lowercase a. Removing the opening and closing brace tokens of the groups and applying the reduction gives us the input of reduction 3. The third reduction is of the form $\backslash K\langle A\rangle\langle B\rangle \mapsto \langle A\rangle$ where both $\langle A\rangle$ and $\langle B\rangle$ are groups. Removing the second group, removing the opening and closing brace tokens of the first group, and applying the reduction gives the input of reduction 4. All remaining reductions are similar except for reductions 9 and 17, which correspond to entering and leaving a group. The last row lists the final result. It is reassuring that the output is 'bac' as expected.

11.5 Defining Commands with TEX

In this section we shall study how to define commands with plain TEX. TEX allows a richer variety of commands than LATEX. The main difference is that TEX commands come in local and global flavours. They may have delimiters in their parameter list and they may be

#	Out	In
1		$\backslash X_1 a_1 b_1 c_1$
2		$\backslash S_1 \ \{_1 \backslash K_2 \{_2 \backslash S_3 \backslash I_3 \}_2 \}_1 \ \{_1 \backslash S_2 \{_2 \backslash K_3 \backslash K_3 \}_2 \backslash I_2 \}_1 \ a_1 b_1 c_1$
3		$\backslash K_1 \{_1 \backslash S_2 \backslash I_2 \}_1 a_1 \{_1 \backslash S_2 \{_2 \backslash K_3 \backslash K_3 \}_2 \backslash I_2 a_2 \}_1 b_1 c_1$
4		$\backslash S_1 \backslash I_1 \{_1 \backslash S_2 \{_2 \backslash K_3 \backslash K_3 \}_2 \backslash I_2 a_2 \}_1 b_1 c_1$
5		$\backslash I_1 b_1 \{_1 \backslash S_2 \{_2 \backslash K_3 \backslash K_3 \}_2 \backslash I_2 a_2 b_2 \}_1 c_1$
6		$\backslash S_1 \backslash K_1 \backslash K_1 b_1 \{_1 \backslash S_2 \{_2 \backslash K_3 \backslash K_3 \}_2 \backslash I_2 a_2 b_2 \}_1 c_1$
7		$\backslash K_1 b_1 \{_1 \backslash K_2 b_2 \}_1 \{_1 \backslash S_2 \{_2 \backslash K_3 \backslash K_3 \}_2 \backslash I_2 a_2 b_2 \}_1 c_1$
8		$b_1 \{_1 \backslash S_2 \{_2 \backslash K_3 \backslash K_3 \}_2 \backslash I_2 a_2 b_2 \}_1 c_1$
9	b	$\{_1 \backslash S_2 \{_2 \backslash K_3 \backslash K_3 \}_2 \backslash I_2 a_2 b_2 \}_1 c_1$
10	b	$\backslash S_2 \{_2 \backslash K_3 \backslash K_3 \}_2 \backslash I_2 a_2 b_2 \}_1 c_1$
11	b	$\backslash K_2 \backslash K_2 a_2 \{_2 \backslash I_3 a_3 \}_2 b_2 \}_1 c_1$
12	b	$\backslash K_2 \{_2 \backslash I_3 a_3 \}_2 b_2 \}_1 c_1$
13	b	$\backslash I_2 a_2 \}_1 c_1$
14	b	$\backslash S_2 \backslash K_2 \backslash K_2 a_2 \}_1 c_1$
15	b	$\backslash K_2 a_2 \{_2 \backslash K_3 a_3 \}_2 \}_1 c_1$
16	b	$a_2 \}_1 c_1$
17	ba	$\}_1 c_1$
18	ba	c_1
	bac	

Table 11.1

TeX's Expansion Processor. The output and the input of the Expansion Processor are listed in the second and third column. The numbers of the reductions are listed in the first column. Each token in the input has a subscript that corresponds to the nesting-level of groups.

defined with and without expanding the substitution text. Usually, you should not need TeX command definitions but sometimes they are needed. The best thing is to define commands using LaTeX commands and only define commands with TeX as a final resort.

The following are TeX's commands for defining commands without delimiters.

`\def⟨cmd⟩#1#2...#n{⟨subst⟩}`

This defines a command, ⟨cmd⟩, with n parameters and with substitution text ⟨subst⟩. The command is local to the group that contains the command's definition. The numbers in the formal parameter list must contain the numbers $1-n$, in increasing order. These restrictions hold for all TeX command definitions. ☑

`\edef⟨cmd⟩#1#2...#n{⟨subst⟩}`

This defines a command, ⟨cmd⟩, with n parameters. The substitution text of the command is the full expansion of ⟨subst⟩. It should be noted that ⟨subst⟩ is expanded when ⟨cmd⟩ is being defined. The command is local to the group that contains its definition. ☑

The following explains the difference between \def and \edef.

```
\def\hi{hi}
\def\hello{\hi}
\edef\ehello{\hi}                    hi. HI.
\def\hi{HI}


\ehello. \hello.
```

The arguments of commands defined with \def or \edef may

not contain paragraph tokens. Paragraph tokens are only allowed in arguments if you add the prefix \long to \def or \edef.

As stated in the explanation of TeX macro definitions, commands may be defined locally in a group. What is more, you may also define macros within other macro definitions. Formal parameters of macro definitions that are nested inside other definitions receive an extra # character to distinguish them from the formal parameters of the nesting macro definition(s). The following is an example.

```
\def\silly#1#2{%
    \def\sillier##1{%
        #2##1#1%
    }%                          James Bond is 007.
    \sillier{#2}%
}
James Bond is \silly70.
```

The following commands are useful for defining low-level commands with TeX.

\csname␣⟨tokens⟩\endcsname

This results in the control sequence of the expansion of ⟨tokens⟩. In effect this expands ⟨tokens⟩ and puts a backslash character to the front of the result. For example, \csname␣command\endcsname gives \command. To see why expansion matters, let's assume we have the definition \def\ho{hoho}. With this definition \csname␣Ho\ho\endcsname gives us \Hohoho. ☑

\noexpand⟨token⟩

This results in ⟨token⟩ without expanding it. For example, the definitions \def\hello{\hi} and \edef\hello{\noexpand\hi} are equivalent regardless of the definition of \hi. ☑

\expandafter⟨token⟩⟨tokens⟩

This expands the first token in ⟨tokens⟩ once (using parameters if required) and inserts ⟨token⟩ before the result. ☑

The command \expandafter is frequently used in combination with \csname to construct definitions with parameterised names. The example in Figure 11.3 demonstrates this mechanism. In this example, the \expandafter lets \csname and \endcsname construct the control sequence name before applying the \def command.

TeX also allows commands with delimiters in parameter lists. For example, it lets you implement a command \command that uses the character | to delimit its two parameters. This lets you apply the command to one and two by writing \command|one|two|. Using TeX you define a command like this as follows.

```
\def\command|#1|#2|{...}
```
LaTeX Usage

More complex delimiters are also allowed. For example, combinations of letters, spaces, and control sequences are valid delimiters, even if the control sequences do not correspond to existing commands. It is also not required that all parameters be delimited or that all delimiters be equal.

```
\documentclass{article}
\def\property#1{%
    \expandafter\def%
    \csname#1\endcsname##1{%
        ##1\ is #1%
    }%
}
\property{brilliant}
\property{excellent}
\begin{document}
    \excellent{\TeX} and
    \brilliant{\LaTeX}.
\end{document}
```

TEX is excellent and LATEX is brilliant.

Figure 11.3

Using the \expandafter command

Figure 11.3

Using the \expandafter command

```
% allow @ in macro names
\makeatletter%
\def\cmd#1{%
  \@ifnextchar[%
    % use the given option
    {\cmd@relay{#1}}%
    % use the default option
    {\cmd@relay{#1}[dflt]}%
}
\def\cmd@relay#1[#2]{...}
% disallow @ in macro names
\makeatother
```

```
\makeatletter
\def\cmd#1{%
  \def\cmd@relay##1[##2]{...}
  \@ifnextchar[%
    {\cmd@relay{#1}}%
    {\cmd@relay{#1}[dflt]}%
}
\makeatother
```

Figure 11.4

Defining commands with default parameters

Figure 11.4 provides two different implementations of a contrived command that has one default parameter. In LATEX terms the example defines a user-defined LATEX command that takes two parameters. The *second* parameter is optional with default dflt.

Let's first study the solution on the left. There are two new aspects to this solution. The first is the use of the commands \makeatletter and \makeatother. After calling the command \makeatletter @ symbols are allowed control sequence names. After calling the command \makeatother @ symbols are no longer allowed in control sequence names. This is a common idiom because it lets you—with high probability—define unique control sequences. The second new aspect is the command \@ifnextchar⟨character⟩⟨first⟩⟨second⟩, which looks ahead to see if the next character is equal to ⟨character⟩ without consuming it. It results in ⟨first⟩ if the next character is ⟨character⟩ and results in ⟨second⟩ otherwise. In the solution on the left the user-defined command \cmd looks ahead to see the token following the first parameter, and passes control to the command \cmd@relay with the proper option.

The solution on the right is similar but it defines the relay command locally. It is recalled that formal parameters of nested macro

Figure 11.5
A sectional unit environment

```
\makeatletter
% Save meaning of old \section command.
\let\old@section=\section
\def\section#1#2{%
  % Define section using old \section command.
  \old@section{#2}
  % Define label for the section.
  \label{#1}
}
\makeatother
```

definitions receive extra # characters. Therefore, the formal parameters of \cmd@relay are now ##1 and ##2. This mechanism should let you refer to both the formal parameters of \cmd and the formal parameters of \cmd@relay inside the substitution text of \cmd@relay.

Candidate delimiters inside matching brace pairs are ignored. For example, let's assume we have the following definition.

```
\def\agoin{ old chap}
\def\hows#1\agoin{How are you #1?}
```
LaTeX Usage

Then \hows{Joe\agoin}\agoin gives 'How are you Joe old chap?'

11.6 Tweaking Existing Commands with \let

This section studies how to tweak existing commands, i. e., redefine an existing command in such a way that the command carries out an additional task. To do this we are going to use TeX's \let command to assign the meaning of the original command to a scratch control sequence. Next we redefine the existing command and refer to the scratch control sequence when we carry out the task that was associated with the original command. In the example in Figure 11.5 we redefine the \section command and force it to take one more parameter, which is the label of the section. The resulting command first uses the original \section command to define the section and next uses the \label command to define the label.

11.7 Using More than Nine Parameters

As mentioned in Section 11.2, LaTeX does not allow more than nine parameters. This section describes two techniques to overcome this problem. Both techniques exploit the fact that TeX macros may have local macro definitions.

To illustrate the solutions we shall implement a command \command that takes ten parameters and outputs their values. The first technique is to implement \command as a wrapper command that does two things.

```
\makeatletter
\def\cmd#1#2#3#4#5#6#7#8#9{%
  \def\cmd@arg@A{#1}%
  \def\cmd@arg@B{#2}%
  ⋮
  \def\cmd@arg@I{#9}%
  \relay%
}
\def\relay#1{%
  Parameters: \cmd@arg@A, \cmd@arg@B, ..., and #1.%
}
\makeatother
```

Figure 11.6
Accessing parameters by defining commands

```
\def\cmd#1#2#3#4#5#6#7#8#9{%
  \def\relay##1{Parameters: #1, #2, ..., and ##1.}%
  \relay%
}
```

Figure 11.7
Accessing parameters with a nested definition

○ It formally defines nine local commands. The *i*-th local command results in the value of the *i*-th parameter of \command.
○ It calls a 'relay' macro that can see the tenth parameter.

Figure 11.6 demonstrates the technique. The second technique is simpler and implements \relay as a local macro. This technique is shown in Figure 11.7.

11.8 Using Environments

This section shows environments and how to define them. The following are a few arguments in favour of environments.

less ambiguity If commands with parameters are used as part of other commands with parameters then this may make it difficult to see which closing brace belongs to which command. If environments are used inside other environments then it is easier to see which \begin{⟨env⟩} belongs to which \end{⟨env⟩}, thereby resolving the brace ambiguity.

more efficiency Environments can be implemented without the need of extra stack space. This makes their implementation more efficient than macros.

The key to defining environments is the command \newenvironment, which is used as follows.

\newenvironment{⟨name⟩}{⟨begin subst⟩}{⟨end subst⟩}
This defines a new global environment called ⟨name⟩. When you write \begin{⟨name⟩}⟨body⟩\end{⟨name⟩} the text ⟨begin subst⟩ is substituted for \begin{⟨name⟩} and the text ⟨end subst⟩ is substituted for \end{{⟨name⟩}}. Effectively, this gives you ⟨begin subst⟩⟨body⟩⟨end subst⟩. ▽

Figure 11.8
User-defined environment

```
\newenvironment{SectionalUnit}[2][section]
                {\csname#1\endcsname{#2}%
                 \begin{refsection}}
                {\printbibliography%
                 \end{refsection}}

\begin{document}
  \begin{SectionalUnit}[chapter]{Introduction}
    \begin{SectionalUnit}{Conventions}
      ...
    \end{SectionalUnit}
    \begin{SectionalUnit}{Notation}
      ...
    \end{SectionalUnit}
  \end{SectionalUnit}
  ⋮
\end{document}
```

\newenvironment{⟨name⟩}[⟨digit⟩]{⟨begin subst⟩}{⟨end subst⟩}

This defines a new global environment ⟨name⟩ with ⟨digit⟩ parameters. In addition to the mechanism for environments without parameters there is now also parameter substitution. However, parameter substitution only works within ⟨begin subst⟩. This works the same as for commands, so the ith actual parameter of the environment is substituted for the ith formal parameter, #i, in ⟨begin subst⟩. It is not allowed to refer to formal parameters in ⟨end subst⟩. ☑

\newenvironment{⟨name⟩}[⟨digit⟩][⟨default⟩]{⟨begin subst⟩}{⟨end subst⟩}

This defines a new global environment called ⟨name⟩ that takes ⟨digit⟩ parameters. The first parameter is optional and has default value ⟨default⟩. ☑

The command \renewenvironment is for redefining environments. It works as expected.

Figure 11.8 presents an example of a user-defined environment that takes two parameters, one of which is optional. The environment defines the start and end of a sectional unit and prints a bibliography at the end of the sectional unit. Such environment definitions are typical: you build more complex environments in terms of existing commands and environments. Of course the environment may not be particularly useful if you don't want a bibliography at the end of your sections.

CHAPTER 12
Branching

THIS CHAPTER IS DEVOTED to decision making and branching. The techniques in this chapter let you implement conditional and iterative statements in LaTeX. Also they let you use or omit text depending on a global mode. This may be useful if you want to generate different output documents from the same source, e.g., a lecture presentation and lecture notes. This gives you ultimate control over the style *and* content of your documents.

12.1 Counters, Switches, and Lengths

This section studies counters, switches, and length-related commands. We study these notions because they play the rôle of variables in LaTeX and TeX.

12.1.1 *Counters*

A LaTeX counter is a *global* variable for counting things. As the name suggests, the values of counters should be integers. Counter names are usually letter sequences. For example, page is a valid name for a counter. The following are the commands related to LaTeX counters.

`\newcounter{⟨name⟩}`
This defines a new global counter, which is a LaTeX variable that can take integer values. It is not quite clear which range is allowed for counters, except that (some) positive, (some) negative, and (all!) zero values are allowed. The initial value of the counter is zero. Lamport [1994, page 138] forbids the use of `\newcounter` in files that are included with the `\include` command. The `\newcounter` command may only be used in the document preamble [Lamport 1994, page 99]   ☑

`\setcounter{⟨name⟩}{⟨value⟩}`
This assigns the value ⟨value⟩ to the counter ⟨name⟩. Here ⟨name⟩ should be the name of an existing counter and ⟨value⟩ should be an integer constant.   ☑

`\stepcounter{⟨name⟩}`
This increments the counter ⟨name⟩ by one. As with `\setcounter`, ⟨name⟩ should be the name of an existing counter.   ☑

`\addtocounter{⟨name⟩}{⟨increment⟩}`
The adds the constant ⟨increment⟩ to the counter ⟨name⟩. As before,

⟨name⟩ should be the name of an existing counter and ⟨increment⟩ should be an integer constant. ☑

\the⟨name⟩

This typesets the value of the counter ⟨name⟩, which should be the name of an existing counter. Here \the⟨name⟩ is the concatenation of \the and ⟨name⟩. For example, the counter section is used in LaTeX for counting the current section number, and the command \thesection gives you the number of the current section. ☑

\newcounter{⟨slave⟩}[⟨master⟩]

This defines a *slave counter* ⟨slave⟩ that depends on a unique *master counter* ⟨master⟩, which should be an existing counter. Slave counters are numbered "within" their master counters. For example, subsection is a slave counter of its master counter section. When ⟨master⟩ is incremented with the \stepcounter command this resets ⟨slave⟩. This version of the \newcounter command is useful for implementing counter hierarchies. ☑

The following example demonstrates these counter-related commands, except for the version of \newcounter with the optional argument.

```
\newcounter{ans}
\setcounter{ans}{9}
\addtocounter{ans}{11}
\stepcounter{ans}
\addtocounter{ans}{\theans}
The answer to the ultimate
 question of life, the universe,
 and everything is \theans.
```

The answer to the ultimate question of life, the universe, and everything is 42.

12.1.2 Switches

LaTeX does not support decision making. To make decisions you need TeX or use a package such as ifthen. In the remainder of this section we shall study TeX decision making commands. The ifthen package is studied in Section 12.2.

\newif\if⟨switch⟩

This is TeX's way to define a branching command called \if⟨switch⟩. We shall refer to such branching commands as *switches*. For example, you may define a switch called \ifnotes with the command \newif\ifnotes. ☑

\⟨switch⟩true

This turns the switch \if⟨switch⟩ on. ☑

\⟨switch⟩false

This turns the switch \if⟨switch⟩ off. ☑

\if⟨switch⟩⟨then clause⟩\fi

This is TeX's equivalent of a conditional one-way branching statement. As expected this results in ⟨then clause⟩ if the switch \if⟨switch⟩ is on. This explanation assumes that ⟨then clause⟩ is expanded. ☑

Unit	Name	Equivalent
pt	point	
pc	pica	1 pc = 12 pt
in	inch	1 in = 72.27 pt
bp	big point	72 bp = 1 in
cm	centimetre	2.54 cm = 1 in
mm	millimetre	10 mm = 1 cm
dd	didôt point	1157 dd = 1238 pt
cc	cicero	1 cc = 12 dd
sp	scaled point	65536 sp = 1 pt

Table 12.1
Length units

\if⟨switch⟩⟨then clause⟩\else⟨else clause⟩\fi

This is the equivalent of a conditional two-way branching statement. It results in ⟨then clause⟩ if the switch \if⟨switch⟩ is on and results in ⟨else clause⟩ otherwise. This explanation assumes that both ⟨then clause⟩ and ⟨else clause⟩ are expanded.   ☑

The following is an example that creates a section. The title of the section depends on the value of the switch \ifnotes. If the switch is on then the title is set to 'Lecture Notes.' Otherwise, the section is titled 'Presentation.' This example can be taken further to implement a context-sensitive document the style *and* content of which depend on the values of switches.

```
\newif\ifnotes
\notestrue

\begin{document}
\section{\ifnotes Lecture Notes%
        \else Presentation%
        \fi}
...
\end{document}
```
LaTeX Usage

The tagging package provides high-level support to do similar things [Longborough 2011].

12.1.3 *Lengths*

This chapter studies length variables, which are LaTeX variables that can be assigned length values. Length variables are also be used for decision making. This section is mainly based on [Lamport 1994, Section 6.4].

LaTeX has a wide range of length (measure) units. Table 12.1 lists them all. Each length unit represents its own length. When LaTeX expects a length, writing 1⟨unit⟩ results in the length of the unit ⟨unit⟩. For example 1mm gives you the length of one millimetre. Likewise you multiply ⟨unit⟩ by any constant ⟨constant⟩ by writing ⟨constant⟩⟨unit⟩. For example, 101in is equivalent to 256.54cm.

Length variables hold length values. You write them just as control sequences. Multiplying length variables works by adding a constant before the variable. For example, 2⟨length⟩ gives you twice the current value of ⟨length⟩.

There are two kinds of lengths: *rigid* and *rubber*. The following explains the difference between the two.

rigid A rigid length always has the same size.

rubber A rubber length is a combination of natural length and elasticity. Rubber lengths may stretch or shrink depending on the situation. This is useful for stretching or shrinking inter-word space and so on. Multiplying a rubber length by a constant results in a rigid length. The result is obtained by multiplying the constant by the natural length of the rubber length. For example, 2.0\rubber gives you twice the natural length of \rubber.

The following are some of LaTeX's length-related commands. By defining your formatting commands in terms of these commands you can make them work regardless of the current document settings.

\parindent

This length variable stores the amount of indentation at the beginning of a normal paragraph. ☑

\textwidth

This length variable stores the width of the text on the page. ☑

\textheight

This length variable stores the height of the body of a page, excluding the head and foot space. ☑

\parskip

This length variable stores the extra vertical space between paragraphs. It is a rubber length with a natural length of zero. A zero natural length usually does not result in additional inter-paragraph spacing. ☑

\baselineskip

This length variable stores the vertical distance between adjacent base lines. ☑

The following are the commands that define and manipulate lengths.

\newlength{⟨command⟩}

This defines the length command ⟨command⟩ with an initial value of 0cm. For example, the command \newlength{\mylen} defines a new length command called \mylen. ☑

\setlength{⟨command⟩}{⟨length⟩}

This assigns the length value ⟨length⟩ to the length command ⟨command⟩. For example, the command \setlength{\parskip}{1.0mm} assigns the value 1mm to \parskip. ☑

\addtolength{⟨command⟩}{⟨length⟩}

This adds the length value ⟨length⟩ to the current value of the length

command ⟨command⟩. For example, the spell \addtolength{\parskip}
{1.0mm} adds a millimetre to \parskip. ☑

\settowidth{⟨command⟩}{⟨stuff⟩}

This assigns the width of ⟨stuff⟩ to ⟨command⟩. For example, the
command \settowidth{\twoms}{MM} assigns the width of the text
'MM' to \twoms. ☑

\settoheight{⟨command⟩}{⟨stuff⟩}

This assigns the height of the bounding box of ⟨stuff⟩ to ⟨command⟩.
For example, the command \settoheight{\tower}{2^{2^2}} assigns the height of 2^{2^2} to \tower. ☑

\settodepth{⟨command⟩}{⟨stuff⟩}

This assigns the depth of the bounding box of ⟨stuff⟩ to ⟨command⟩.
For example, the command \settodepth{\depth}{amazing} sets the
value of \depth to the distance that the letter g extends below the base
line. ☑

The commands \setlength and \addtolength obey the normal
scoping rules.

12.1.4 *Scoping*

This section briefly explains the difference between the scoping rules
for assignments to counters, TeX switches, and lengths. Counters
are *global,* which is to say that the values of counter variables are *not*
restored upon leaving the group. TeX switches and LaTeX lengths satisfy
group scoping rules, which means that these variables are restored to
the same values that they had when the group was entered.

12.2 The ifthen Package

The ifthen package provides Boolean variables at the LaTeX level, decision making, and branching. There are two commands for defining
new Boolean variables.

\newboolean{⟨bool⟩}

This defines a new global Boolean variable. the command will fail if
⟨bool⟩ is already defined. ☑

\provideboolean{⟨bool⟩}

This also defines a new global Boolean variable. However, this command will accept ⟨bool⟩ if it is already defined. ☑

\setboolean{⟨bool⟩}{⟨value⟩}

This assigns the value ⟨value⟩ to ⟨bool⟩. Here ⟨value⟩ should be true
or false. ☑

Knowing how to define Boolean variables we can proceed with
decision making.

\ifthenelse{⟨test⟩}{⟨then clause⟩}{⟨else clause⟩}

This command is a two-way branching construct. As expected it carries
out ⟨then clause⟩ if ⟨test⟩ evaluates to true and carries out ⟨else

clause⟩ if ⟨test⟩ evaluates to false. The condition ⟨test⟩ must be a
valid condition.   ☑

Valid conditions for the ⟨test⟩ argument of the \ifthenelse command are as follows.

⟨**boolean**⟩

A Boolean constant that should be true or false, ignoring case, so
true, truE, ..., TRUe, and TRUE are equivalent, and so are false, falsE,
..., FALSe, and FALSE.   ☑

⟨**number₁**⟩⟨**op**⟩⟨**number₂**⟩

Here ⟨number₁⟩ and ⟨number₂⟩ should be numbers and ⟨op⟩ should be
<, =, or >.   ☑

lengthtest{⟨**dimen₁**⟩⟨**op**⟩⟨**dimen₂**⟩}

Here ⟨dimen₁⟩ and ⟨dimen₂⟩ should be dimension values and ⟨op⟩
should be <, =, or >.   ☑

isodd{⟨**number**⟩}

As suggested by the notation ⟨number⟩ should be a number.   ☑

isundefined{⟨**command**⟩}

Here ⟨command⟩ should be a control sequence name.   ☑

equal{⟨**string₁**⟩}{⟨**string₂**⟩}

Here ⟨string₁⟩ and ⟨string₂⟩ are evaluated and compared for equality.
The test is equivalent to true if and only if the results of the evaluations
are equal.   ☑

boolean{⟨**bool**⟩}

Here ⟨bool⟩ should be a Boolean variable.   ☑

⟨**test₁**⟩⟨**command**⟩⟨**test₂**⟩

Here ⟨test₁⟩ and ⟨test₂⟩ should be valid ⟨test⟩ conditions and ⟨command⟩
should be \or, \and, \OR, or \AND. The versions \OR and \AND are preferred to \or and \and because they are more robust.   ☑

⟨**negation**⟩⟨**test**⟩

Here ⟨test⟩ should be a valid ⟨test⟩ condition and ⟨negation⟩ should
be \not or \NOT. The uppercase version is preferred to the lowercase
version.   ☑

(⟨**test**⟩**\)**

Here ⟨test⟩ should be a valid ⟨test⟩ condition.   ☑

The following example demonstrates how to use the \ifthenelse
command. The page counter variable that is used in the example keeps
track of LaTeX's page numbers. It is assumed the ifthen package is
loaded.

```
\begin{document}
    \ifthenelse
        {\isodd{\value{page}}}          Odd page.
        {Odd page.}
        {Even page.}
\end{document}
```

whiledo{⟨**test**⟩}{⟨**statement**⟩}

The command \whiledo{⟨test⟩}{⟨statement⟩} is ifthen's equivalent of the while statement. It repeatedly 'executes' ⟨statement⟩ until
⟨test⟩ becomes false. The following example demonstrates some

of the functionality of the `ifthen` package. It is assumed the `ifthen` package is loaded.

```
\newcounter{cnt}
\setcounter{cnt}{3}
$\thecnt =
 \whiledo
    {\not\(\thecnt=0\)}%
    {+ 1 \addtocounter{cnt}{-1}}$.
```
$3 = 0 + 1 + 1 + 1.$ ☑

12.3 The `calc` Package

The `calc` package extends TeX and LaTeX's arithmetic. The `calc` package redefines the commands \setcounter, \addtocounter, \setlength, and \addtolength. As a result, these commands now accept infix expressions in their arguments. The package also provides useful commands such as \widthof{⟨stuff⟩}, \ratio{⟨dividend⟩}{⟨divisor⟩}, and so on, which don't have a LaTeX equivalent. The interested reader is referred to the package's excellent documentation [Krab Thorub, Jensen, and Rowley 2007].

12.4 Looping

The LaTeX kernel provides two kinds of `for` statements.

\@for \var:=⟨list⟩\do \command

Here ⟨list⟩ is a comma-delimited list. The items in ⟨list⟩ are bound to \var from left to right. After each binding, the command \command is carried out. (Of course, \command can also be a group.) The following is an example. Note that it is assumed that the symbol @ is allowed in control sequence names.

```
\@for \var:=1,two\do{%
    (\var)%
}
```
(1)(two) ☑

\@tfor\var :=⟨list⟩\do \command

This is the "token" version of the \@for command. In this case ⟨list⟩ is a list of tokens. The tokens in ⟨list⟩ are bound to \var from left to right. After each binding, the command \command is carried out.

```
\newcommand*\swop[2]{#2#1}
\@tfor\var:=1\swop\do{%
    \var23%
}
```
12332 ☑

The LaTeX kernel also provides the following `while` statement.

\@whilesw⟨switch⟩\fi{⟨statements⟩}

This is a `while` loop with a condition that is based on a TeX switch ⟨switch⟩. The following is an example.

Figure 12.1

Tail recursion. The command \apply \cmd ⟨items⟩ \endApply applies \cmd to each item in ⟨items⟩. For example, \apply\twice a{bc}\endApply gives aabcbc. The key to understanding the macro is noticing that \breakApply substitutes \fi for all tokens up to and including \Apply.

```
\def\apply#1{%
   \def\Apply##1{%
      \ifx##1\endApply%
         \breakApply% terminate recursion
      \fi%
      #1{##1}% Apply command to next item.
      \Apply%  Tail recursive call.
   }%
   \Apply%
}
\def\breakApply#1\Apply{\fi}%
\def\twice#1{#1#1}

\apply\twice a{bc}d\endApply
```

```
\newif\iffirst\firsttrue
\newif\ifsecond\secondfalse
\@whilesw\iffirst\fi{%
  X\ifsecond\firstfalse%
      \else\secondtrue\fi%
}
```
                                    XX   ☑

Scharrer [2011-07-23] provides an interesting list of more low-level LaTeX kernel commands.

12.5 Tail Recursion

In computer science a function is called tail recursive if the function carries out no more than one recursive call in the current incarnation. Tail recursion is a common technique to implement iteration. The remainder of this section shows how you may implement a tail recursive command using low-level TeX delimited macros. After carefully studying this section the interested reader should fully appreciate TeX and LaTeX programming in the large.

The evaluation of the example in Figure 12.1, demonstrates TeX expansion in its full glory. The example is based on [Fine 1992]. There is one new ingredient in the example, which is related to decision making. The construct \ifx⟨A⟩⟨B⟩⟨statement⟩\fi in the example results in ⟨statement⟩ if the tokens ⟨A⟩ and ⟨B⟩ are equal. The key to understanding the example is observing that (1) \breakApply is applied only once inside \Apply, (2) that it is only applied when the token \endapply is detected, and (3) that \breakApply substitutes \fi for the tokens up to and including the first next \Apply. The substitution closes the current \if with a matching \fi, thereby terminating the recursion. The rest all boils down to tail recursion. It is left to the reader to determine the resulting output.

CHAPTER 13
Option Parsing

THIS CHAPTER STUDIES ⟨key⟩=⟨value⟩ interfaces. Such interfaces may be used to implement macros whose parameters are specified as a list of ⟨key⟩=⟨value⟩ pairs. Usually the pairs are provided in the optional argument.

At the time of writing the most frequently-used package for implementing ⟨key⟩=⟨value⟩ interfaces is keyval [Carlisle 1999b]. A more recent package is pgfkeys [Tantau 2010]. To keep this chapter simple, it only discusses pgfkeys because it is more versatile and provides more robust ⟨key⟩=⟨value⟩ parsing.

13.1 What is a ⟨Key⟩=⟨Value⟩ Interface?

In a traditional Application Programming Interface (API), the actual parameters are related to the formal parameters by *positional association*. This means that the position of the actual parameter in the parameter list determines the corresponding formal parameter. A more recently developed API uses *named association*. An API that uses named association tags each actual parameter with the name of its corresponding formal parameter. Usually this is done by providing a list of ⟨key⟩=⟨value⟩ pairs. Each ⟨key⟩ is a formal parameter name and each ⟨value⟩ is the actual parameter.

A ⟨key⟩=⟨value⟩ interface, is LaTeX speak for an API that uses named association to relate actual and formal parameters. The actual parameters (the values) of a macro are specified as a list of ⟨key⟩=⟨value⟩ pairs in the optional argument of a command. For example, the \includegraphics command, which is provided by the graphicx package, inserts external pictures. The required argument of the command is the name of the picture. The command also has an optional argument that lets you specify how the picture should be inserted. This is done by letting you specify the width, the height, the angle of rotation, and many other settings. Each setting is specified as a ⟨key⟩=⟨value⟩ pair. For example, \includegraphics[width=9cm,height=3cm]{pic.png} is a request to insert the picture pic.png with a value of 9 cm for a (formal) parameter called width, and a value of 3 cm for a (formal) parameter called height. Note that the command is flexible because it is not necessary to provide all possible settings. For example, the rotation isn't explicitly specified.

13.2 Why Use a ⟨Key⟩=⟨Value⟩ Interface?

Chapter 11 demonstrated that LaTeX's API is not ideal. The following are some arguments in favour of ⟨key⟩=⟨value⟩ interfaces.

number of arguments There is no limit to the number of ⟨key⟩=⟨value⟩ pairs. In contrast, the standard LaTeX API is limited to 9 parameters.

robustness A ⟨key⟩=⟨value⟩ API is more robust than an API that uses positional association. For example, the order of the ⟨key⟩=⟨value⟩ pairs is irrelevant. Furthermore, default values may be defined for missing ⟨key⟩=⟨value⟩ pairs.

simplicity By relating an actual parameter—a value—to a key, the purpose of the parameter becomes more clear. This makes the API simpler and easier to use.

self-documentation This is related to the previous item. The ⟨key⟩=⟨value⟩ paradigm avoids references to the meaningless positional parameter names #1, #2, ... Instead the programmer can use the more meaningful names of the keys. This makes the code more self-documenting, which makes it easier to reason about the implementation. As a consequence this reduces the possibility of errors.

13.3 The pgfkeys Package

The pgfkeys package [Tantau 2010] is a recent alternative to the key-val package [Carlisle 1999b]. As the name suggests, the package is implemented by the makers of pgf. The package implements very robust ⟨key⟩=⟨value⟩ parsing. For example, if there aren't any commas in a value then there's usually no need to put the value inside braces. Furthermore, the package allows key hierarchies with keys consisting of one or several words. These hierarchies may be viewed as a Unix file system. The package also generalises ⟨key⟩=⟨value⟩ parsing by allowing multiple values for a given key. However, this feature is usually not needed for simple ⟨key⟩=⟨value⟩ interfaces. Finally, Wright [2010] introduces the pgfopts package, which provides pgfkeys-style option parsing for class and package options. We shall use the pgfopts package in Chapter 15.

The remainder of this chapter discusses a selection of pgfkeys techniques that should be enough for day-to-day use. Specifically, we shall study providing and using values of keys, traversing the key tree, error handling, storing values in macros, decision keys, and choice keys. The pgfkeys package provides many more techniques but they are beyond the scope of this book. You may find more about the package by reading the package documentation, which is contained in the pgf manual [Tantau 2010].

13.4 Providing and Using the Values

The most common reason for using a ⟨key⟩=⟨value⟩ interface is to let users provide a value for a given key and to record the value for that key. This is usually done by defining a command that records the value of a given user-provided ⟨key⟩=⟨value⟩ pair.

In this section we shall start by studying how to provide values for the keys and how to use these values. Using the values in this section only results in output and has no other side-effects. In Section 13.8 we shall study how to record the values.

\pgfkeys{⟨key⟩/.code=⟨expr⟩}

This defines the code of the key ⟨key⟩. The code is best viewed as a zero or one-parameter macro with substitution text ⟨expr⟩. When the user provides a value for the key, the key is substituted for the positional parameter #1 in ⟨expr⟩.

The following shows how this works. The first line defines the code for the key /greeting, and the next two lines provide values for the key.

```
\pgfkeys{/greeting/.code=Hello #1.}        Hello moon.
\pgfkeys{/greeting=moon}                   Hello world.    ☑
\pgfkeys{/greeting=world}
```

\pgfkeys{⟨key⟩/.default=⟨default⟩}

This defines ⟨default⟩ as the default value for the key ⟨key⟩. If you define ⟨default⟩ as the default value for ⟨key⟩ then this makes using ⟨key⟩ without a value equivalent to executing \pgfkeys{⟨key⟩=⟨default⟩}. The following continues the previous example.

```
\pgfkeys{/greeting/.default=sun}           Hello stars.
\pgfkeys{/greeting=stars}                  Hello sun.    ☑
\pgfkeys{/greeting}
```

13.5 Traversing the Key Tree

As pointed out earlier on, pgfkeys keys may be regarded as paths in a key tree that is similar to a Unix file system. The tree's root node is indicated with a forward slash (/). You may create a child of a parent node by appending an extra forward slash to the name of the parent and by adding the name of the child after the extra forward slash. For the root we usually leave out the extra forward slash. Forward slashes at the end of the names are not significant. For example, the root's child that is called solar system is denoted /solar system. The child called earth of the node /solar system is denoted /solar system/earth, and so on.

Paths with names starting with a forward slash are called *absolute paths*. Needless to say, referring to paths by their absolute names is tedious and prone to errors. This is one of the reasons why Unix also has relative paths. Relative paths don't start with a forward slash and they're relative to a working directory—also known as the current directory. In Unix you change to a different working directory with the cd command. When you change to a new working directory all relative paths become relative to the new working directory. The pgfkeys package provides a similar functionality. Relative key paths are key paths that are relative to the current key path. You can choose a new current (key) path by adding the text /.cd to the name of that path. Just as in Unix, you are free to use relative or absolute path notation. The following explains how to change the current path.

\pgfkeys{⟨path⟩/.cd,⟨stuff⟩}

This makes ⟨path⟩ the current path in ⟨stuff⟩. The next example shows how this works. We first make /cork/greeting the current path (key). Next we define the default value and the code for the key. Finally, we use the key.

```
\pgfkeys{/cork/greeting/.cd,
         .default=boie,
         .code=Howsagoin #1.}
\pgfkeys{/cork/greeting=Liz,
         /cork/greeting}
```

Howsagoin Liz.
Howsagoin boie.☑

13.6 Executing Keys

A key that executes one or several keys is called a style.

\pgfkeys{⟨key⟩/.style=⟨list⟩}

This defines ⟨list⟩ as a style for ⟨key⟩. If ⟨list⟩ contains several keys then it should be enclosed in braces. When a user uses ⟨key⟩, this results in ⟨list⟩. If the positional parameter #1 occurs in ⟨list⟩ then the current value of ⟨key⟩ is substituted for the parameter.

The following continues our running example. In the example we define a new style called /cork/greetings. The style executes two keys. The first executed key is /cork/greeting=Mr~⟨value⟩, where ⟨value⟩ is the value of the new style. The second executed key is /cork/greeting.

```
\pgfkeys{/cork/greetings/.style={
         /cork/.cd,
         greeting=#1,
         greeting}}
\pgfkeys{/cork/greetings=Roy}
```

Howsagoin Roy.
Howsagoin boie.☑

13.7 Error Handling

Adding error handling is good practice because it detects errors and helps to diagnose them. We've already seen the .default qualifier, which defines a default value for its key. The following qualifiers state which keys require a value and which keys don't take values.

\pgfkeys{⟨key⟩/.value required}

This makes a value mandatory when ⟨key⟩ is used. The value is required even if ⟨key⟩ has a default value.

```
\pgfkeys{/homer/drink/.cd,
        .code=#1,
        .value required}                    beer            ☑
\pgfkeys{/homer/drink=beer}
\pgfkeys{/homer/drink}% D'oh
```

\pgfkeys{⟨key⟩/.value forbidden}

This forbids values when ⟨key⟩ is used. Providing a value results in an error.

```
\pgfkeys{/homer/lunch/.cd,
        .code=donuts,
        .value forbidden}                   donuts          ☑
\pgfkeys{/homer/lunch}
\pgfkeys{/homer/lunch=peas}% D'oh
```

13.8 Storing Values in Macros

The most common application of ⟨key⟩=⟨value⟩ interfaces is to store the value of a given key. The following shows how to do this.

\pgfkeys{⟨key⟩/.store in=⟨command⟩}

This stores the value of ⟨key⟩ in the command ⟨command⟩. The value is not expanded. Note that if you define store a value for a key then the code of the key is no longer used. The following is an example.

```
\newcommand*\a{a}
\pgfkeys{/storage/.store in=\myget}
\pgfkeys{/storage=a is \a.}          Before: a is a.
Before: \myget                       After: a is A.    ☑
\renewcommand*\a{A}
After:  \myget
```

\pgfkeys{⟨key⟩/.estore in=⟨command⟩}

This works as .store in but it expands the value before saving it in ⟨command⟩.   ☑

13.9 Decisions

Letting users turn options on and off is another common application of ⟨key⟩=⟨value⟩ interfaces. The pgfkeys package implements such options as user-defined *decision keys*. Decision keys take Boolean values. Decision keys are used in combination with TeX switches that reflect the values of the decision keys.

\pgfkeys{⟨key⟩/.is if=⟨switch⟩}

This defines ⟨key⟩ as a decision key with a TeX switch \if⟨switch⟩. Valid values for decision keys are true and false. The default value is true. It is the user's responsibility to define the TeX switch.

```
\newif{\ifswitch}
\pgfkeys{/decision/.is if=switch}
\pgfkeys{/decision}
\ifswitch ON\else OFF\fi.                    ON.
\pgfkeys{/decision=false}                     OFF.
\ifswitch ON\else OFF\fi.                     ON.           ☑
\pgfkeys{/decision=true}
\ifswitch ON\else OFF\fi.
```

13.10 Choice Keys

Our final pgfkeys application is *choice keys,* which are keys that can take values from a predefined list of values.

\pgfkeys{⟨key⟩/.is choice}

This makes ⟨key⟩ a choice key. By defining a style for ⟨key⟩/⟨option⟩ you make value ⟨option⟩ a valid value for ⟨key⟩. When the user selects ⟨option⟩ as a value for ⟨key⟩, this executes the style ⟨key⟩/⟨option⟩. The following example should explain the mechanism. The example results in '1 3 2.'

```
\newcommand*\mycount{0}                              LATEX Input
\pgfkeys{/counter/.store in=\mycount}

\pgfkeys{/selection/.cd,
        .is choice,
        first/.style={/counter=1},
        second/.style={/counter=2},
        third/.style={/counter=3}}
\pgfkeys{/selection=first}  \mycount
\pgfkeys{/selection=third}  \mycount
\pgfkeys{/selection=second} \mycount                          ☑
```

PART VI

Miscellany

Oil and charcoal on canvas (06/10/10 no 2), 64 × 91 cm
Work included courtesy of Billy Foley
© Billy Foley (www.billyfoley.com)

CHAPTER 14
Beamer Presentations

THIS CHAPTER INTRODUCES the beamer class, which is widely used for computer presentations. Some people call such presentations *powerpoint presentations*. The beamer class is seamlessly integrated with the tikz package and lets you present *incremental* presentations, which are presentations that incrementally add text and graphics to a page of the presentation.

The purpose of this chapter is *not* to explain all the possibilities of the beamer class but to explain just enough for what is needed for one or two presentations. The interested reader is referred to the excellent documentation [Tantau, Wright, and Miletić 2010] for further information.

The remainder of this chapter is as follows. In Section 14.1 we shall study *frames*, which correspond to one or several incremental slides on the screen. Section 14.2 explains the concept of *modal* presentations, which let you generate different versions of your presentation. For example, an in-class presentation and a set of lecture notes. This is continued in Section 14.3, which studies incremental presentations. Section 14.4 shows how to add some visual "alert" effects. This may be useful for highlighting certain parts of the presentation. This chapter concludes with Section 14.5, which spends a few words on how you may personalise your presentations by adding a dash of style.

14.1 Frames

The frame environment is to a computer presentation what a page is to an article, a report, or a book. However, a frame may also be decorated with a frame title and a frame subtitle. Throughout this section we shall not worry about the overall look and feel of the presentation.

`\begin{frame}[options]` ⟨frame material⟩ `\end{frame}`

This is a simplified presentation of the frame environment (Section 14.2 provides a more complete description). When the output document is a computer presentation the ⟨frame material⟩ is turned into one or several slides in the output. Otherwise, it may result in one or several lines of text in the text of your output document.

If the option fragile is included in ⟨options⟩, then ⟨frame material⟩ may contain any LaTeX material. Including the option fragile is by far the easier: just use it. Omitting the fragile option may result in errors caused by so-called "verbatim" commands and environments.

Figure 14.1

Creating a titlepage with the beamer class. The outline of the output slide is drawn for clarity. The little pictures in the lower right corner of the output are for navigation purposes.

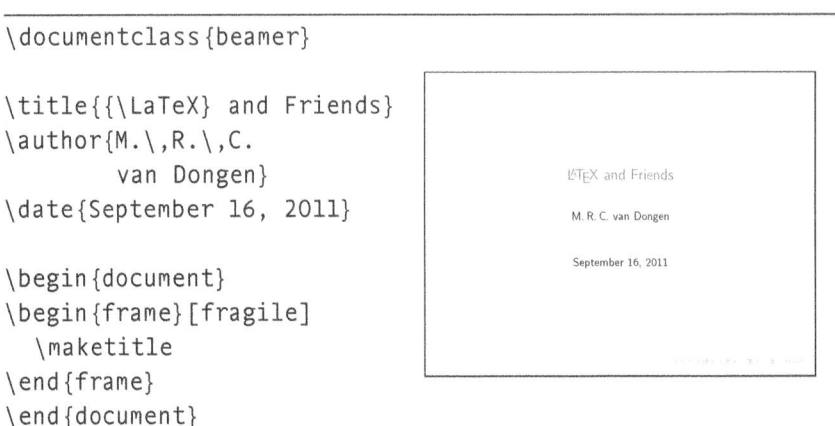

```
\documentclass{beamer}

\title{{\LaTeX} and Friends}
\author{M.\,R.\,C.
        van Dongen}
\date{September 16, 2011}

\begin{document}
\begin{frame}[fragile]
  \maketitle
\end{frame}
\end{document}
```

Figure 14.2

Creating frame titles. The outline of the output slide is drawn for clarity.

```
\begin{frame}[fragile]
  \frametitle{A Slide}
  \framesubtitle{An Example}

  \begin{itemize}
  \item Hello world.
  \item Bonjour monde.
  \end{itemize}
\end{frame}
```

Tantau, Wright, and Miletić [2010, Chapter 8] provides further information about the fragile option.

The following is important: the \begin{frame} and \end{frame} commands should be on a line of their own and there should be no spaces before the \begin and \end.

Figure 14.1 provides the first beamer example. As you may see from the example, it looks like a regular LaTeX document with a \title, \author, and \maketitle command. However, since beamer is a document class, its name is included in the \documentclass argument. The command \maketitle is put in a frame (environment). ☑

\frametitle{⟨frame title⟩}

This defines a frame title, which is usually typeset at the top of the resulting slides of a computer presentation. The frame title is only included if the output document is text-based. However, as we shall see in Section 14.2 it is possible to turn the frame title off for such documents. Turning the frame title off is also possible by redefining the \frametitle command. ☑

\framesubtitle{⟨frame subtitle⟩}

This defines a subtitle for the frame. The subtitle is usually typeset below the frame title. ☑

Figure 14.2 demonstrates a simple beamer frame. The frame has a frame title and subtitle and its body consists of an itemised list.

The beamer class is nice when it works but it may lead to unexpected complications. For example, the following may not work.

```
\newenvironment{myframe}[0]
                {\begin{frame}[fragile]}
                {\end{frame}}
```

Don't do this at Home

Explaining why this environment doesn't work is beyond the scope of this chapter. As a general rule, automating beamer commands may not always work: don't try it unless you have time. The manual [Tantau, Wright, and Miletić 2010] is the ultimate source of information for what is and isn't possible.

14.2 Modal Presentations

This section shows how to exploit beamer's *modes*, which let you generate several kinds of output documents from the same source. Here a different output document may have a different style of presentation but also different content. The following are beamer's basic modes.

beamer This is the default mode, which is what beamer is "in." It corresponds to a computer presentation with one or several slides per frame. For example, in a presentation you can uncover an itemised list item by item. So the single itemised list in the single frame gives you several slides in the output. In a different mode, beamer may not uncover the itemised list item by item but present the itemised list as a whole.

second This mode is for outputting material to a second output screen.

handout This mode is for handouts. When a frame in the input is typeset in this mode, beamer suppresses uncovering effects and presents the frame as a whole. This is different from the default mode, where one input frame may result in several output slides.

trans This mode is for creating transparancies. Having such an option almost seems like an anachronism. However, having a presentation in the form of transparancies may be useful as a backup resource, e.g., when presenting away from home.

article This mode is for typesetting text using a different existing LaTeX class. For example, this book was typeset using LaTeX's book class in beamer's article mode. Uncovering is suppressed in article mode. Using article mode requires a slightly different approach. This time, you use the \documentclass to load the different class and use the \usepackage command to import the beamerarticle package. Figure 14.3 demonstrates how you may do this. In this example, all frame titles and frame subtitles are turned off by redefining the \frametitle and \framesubtitle commands.

The beamer class is always in one of these five modes. By providing the mode as an optional argument to the beamer class you determine the mode. If you omit the mode then beamer will be in, well, beamer mode. The beamer class also has the following auxiliary modes:

all This is for all modes.

presentation This mode is for all "presentation" modes, so all modes except for article.

Figure 14.3

Using the beamerarticle package

```
\documentclass{book}
\usepackage{beamerarticle}
\makeatletter
\def\frametitle{%
    \@ifnextchar<%
        {\@frametitle@lt}%
        {\@frametitle@lt<>}%
}
\def\@frametitle@lt<#1>#2{}
\makeatother
```

Figure 14.4

Using modes. The outline of the slide is drawn for clarity.

```
\documentclass[handout]
                {beamer}

\begin{document}
\begin{frame}
        <handout|beamer>
        [fragile]
    Beamer or handout mode.
\end{frame}
\begin{frame}
        <beamer>
        [fragile]
    Beamer mode.
\end{frame}
\end{document}
```

Handout or beamer mode.

Now that we know about beamer's modes, it's time to revisit its frame environment.

\begin{frame}<⟨overlay specs⟩>[⟨options⟩] ⟨frame material⟩ \end{frame}

Here <⟨overlay specs⟩> behaves as an optional argument. The modes in ⟨overlay specs⟩ determine whether the frame should be typeset. For example, if ⟨overlay specs⟩ is article and beamer is in beamer mode then the frame is not typeset. You may combine modes using the pipe symbol (|) as a separator. For example, if you use beamer|handout then the frame is typeset if beamer is in beamer or handout mode. ☑

Figure 14.4 demonstrates the basic mode mechanism. The input defines two frame. The first frame is typeset in handout or beamer mode. The second frame is only typeset in beamer mode. The beamer class is started in handout mode. This explains why only the first frame is typeset.

Other beamer commands and environment may also accept overlay specifications. Having to specify the same overlay specification is tedious and prone to errors. The following commands help avoiding redundant overlay specifications.

\mode<⟨mode spec⟩>{⟨text⟩}

This inserts ⟨text⟩ if beamer's mode corresponds to ⟨mode spec⟩. Note that this only works if the first non-space character following the > is an opening brace. ☑

`\mode<⟨mode spec⟩>`

This filters subsequent text that does not correspond to ⟨mode spec⟩. Note that this only works if the first non-space character following the > is *not* a brace. ☑

`\mode*`

When `beamer` is in `presentation` mode, then this command causes `beamer` to ignore text outside `frame` environments. When `beamer` is in `article` mode, this command has no effect. ☑

14.3 Incremental Presentations

Incremental presentations incrementally unveil the content of a `frame` environment. Typically, this is done by displaying the next item in an itemised list. The `beamer` class also provides annotations for presenting material on the nth output slide of a given `frame`. The following are some of the relevant commands. More information may be found in [Tantau, Wright, and Miletić 2010, Chapter 9].

`\pause`

This inserts a pause stop at the corresponding position. In `presentation` mode the command adds one more output slide to the slides that are generated from the current `frame`. The pause stop separates the material before and after the position of the `\pause` command. The slides at the start are unaware of the material after the `\pause` command. For example, assume that you have a `frame` consisting of the text 'hide `\pause` and `\pause` seek.' In `presentation` mode this will result in three slides. The first contains the text 'hide,' the second the text 'hide and,' and the third the text 'hide and seek.' ☑

`\pause[⟨number⟩]`

This command unveils the material following the `\pause` command from slide ⟨number⟩ and onwards. Hardcoding numbers in `\pause` commands like this does not make for maintainable code. For example, let's assume you have nine `\pause` commands `\pause[2]`–`\pause[10]`. If you you want to join the material on Slides 1 and 2 then you have to remove the command `\pause[2]` and renumber the arguments of the remaining eight commands. ☑

Figure 14.5 provides an example of the `\pause` command. The input `frame` results in three output slides. The first slide contains the first item of the itemised list. The second slide contains the first, the second, and the third item. The last slide contains all items of the itemised list. It is assumed that `beamer` is in `beamer` mode.

The `beamer` class redefines the standard `\item` command. The redefined version of the command takes an additional optional argument that acts as an *overlay specification*. The overlay specification determines which slides should contain which items. The optional argument is passed in angular brackets (< and >). Without the overlay specification, the `\item` command works as per usual. Overlay specifications are ignored if `beamer` is not in `presentation` mode. However,

Figure 14.5

Using the \pause command. The frame environment results in three output slides, the second of which is shown on the right. The outline of the slide is drawn for clarity.

```
\begin{frame}[fragile]
\begin{itemize}
\item First. \pause
\item Second.
\item Third. \pause
\item Last.
\end{itemize}
```

when the overlay specification is present in presentation mode, then the material in the scope of the \item is only displayed on the slides corresponding to an overlay specification.

\item<⟨overlay spec⟩>

The corresponding item is typeset on the slides corresponding to ⟨overlay spec⟩. On the remaining slides, the item is typeset in invisible ink. ☑

The following are some possible overlay specifications

⟨number⟩   This corresponds to slide ⟨number⟩.

⟨number⟩-   This corresponds to slide ⟨number⟩ and onward.

-⟨number⟩   This corresponds to slides 1–⟨number⟩.

⟨number₁⟩-⟨number₂⟩   This corresponds to slides ⟨number₁⟩–⟨number₂⟩.

⟨overlay spec₁⟩,⟨overlay spec₂⟩   This combines specifications ⟨overlay spec₁⟩ and ⟨overlay spec₂⟩.

Other commands may also accept overlay specifications. The reader is referred to the class documentation [Tantau, Wright, and Miletić 2010] for further information.

Intermezzo. The beamer class defines many more commands for creating incremental presentations. Incremental presentations may look slick, but creating them takes precious time. Peyton Jones, Huges, and Launchberry [1993] argue that some of your audience may not even like incremental presentations that unveil an itemised list one item at a time. It is the content of the presentation that determines the quality—not the visual effects. As a student you probably won't have to give many presentations. Consider doing yourself and your audience a favour: minimise the visual effects and spend the time you save on the content of the presentation.

The tikz package and the beamer class are seamlessly integrated. This means you can also create incremental presentations with tikz pictures. Such presentations may be highly effective. However, creating them may take a *lot* of time. If you're a student and you only have to present a few presentations, you may be better off by staying away from incremental presentations.

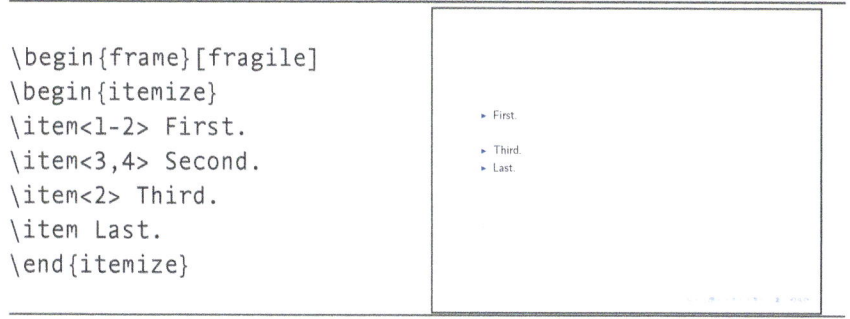

Figure 14.6
Using overlay specifications. The `frame` environment results in three output slides, the second of which is shown on the right. The outline of the slide is drawn for clarity.

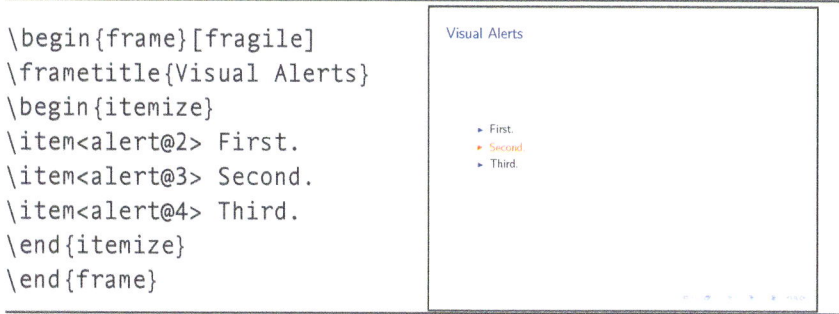

Figure 14.7
Adding visual alerts. The `frame` to the left results in four output slides. The first slide has no visual alerts. The remaining slides highlight the items in the list. The third slide is shown on the right. The second alert in the input draws the second output item in red. The outline of the slide is drawn for clarity.

14.4 Visual Alerts

A visual alert in a presentation uses colour to emphasise text. Using visual alerts is useful if you want to emphasise different parts of a `frame` at different times. It is especially useful if you're discussing items in a list and if you want to indicate which item is currently being discussed. The following are some related commands.

`\alert<⟨overlay spec⟩>{⟨text⟩}`
This emphasises ⟨text⟩ on the slides corresponding to ⟨overlay spec⟩. Omitting ⟨overlay spec⟩ results in highlighting ⟨text⟩ on all slides. ☑

`\item<alert@⟨overlay spec⟩>`
This emphasises the current item in a list on the slides corresponding to ⟨overlay spec⟩. ☑

`\item<⟨overlay spec₁⟩|alert@⟨overlay spec₂⟩>`
This displays the current item on the slides corresponding to ⟨overlay spec$_1$⟩ and emphasises the item on the slides corresponding to ⟨overlay overlay spec$_2$⟩. ☑

Figure 14.7 uses visual alerts to highlight the different items in an itemised list.

14.5 Adding Some Style

The presentation in this chapter has been quite minimal because learning the `beamer` class takes time. Students should use simple presentation styles and spend their time on the presentation's content.

Having made these observations, it is good to note that some

232 | Chapter 14

presentations benefit from some additional decoration. For example, a menu that lists the sections in the presentation may help the audience recognise the structure of the presentation.

A `beamer` *theme* determines a certain aspect of the visual presentation. Currently, there are five `beamer` themes: presentation, colour, font, inner, and outer. The presentation themes are the easier ones to use because they define *everything* in the presentation. New `beamer` users are better off starting with a presentation theme because then they don't have worry about the presentation style. Most presentation themes are actually quite good. Seasoned `beamer` users may want to spend some time on fine-tuning their own style.

The remainder of this section presents four presentation themes that are ideal for a first presentation with only a few slides. More information about themes may be found in the documentation [Tantau, Wright, and Miletić 2010].

The input that was used to demonstrate the different themes in included in Figure 14.12 at the end of this chapter. The input is inspired by the `beamer` documentation. The resulting outputs are listed in Figures 14.8–14.11. For each theme, the figure contains the fifth slide, i. e., the fourth slide of the second `frame`.

Figure 14.8 depicts the sample output of `beamer`'s `default` theme. This theme is very sober and implements visual alerts by typesetting text in red, which is the default for visual alerts.

The `Boadilla` theme, which is depicted in Figure 14.9, is a bit more lively. Using this theme also adds some information about the "author" at the bottom of each slide. Passing the option `secheader` also lists the current section and subsection at the bottom of the slides.

Sample output of the `Antibes` theme is depicted in Figure 14.10. This theme adds a tree-line navigation menu to the `Boadilla` theme. This kind of information may be useful for the audience because it helps them recognise the presentation stucture and helps them determine where you "are" in the presentation.

The final theme is `Goettingen`. It is depicted in Figure 14.11. This theme is for long presentations and comes equipped with a table of contents in a sidebar. This theme accepts the following options.

`left`
This puts the sidebar on the left of the screen. ☑

`right`
This puts the sidebar on the right of the screen. This is the default behaviour. ☑

`width=⟨dimension⟩`
This sets the width of the sidebar. Providing a width of zero hides the sidebar. ☑

`hideallsubsections`
This removes subsection information from the sidebar. ☑

`hideothersubsections`
With this option only the subsections of the current section are shown in the sidebar. ☑

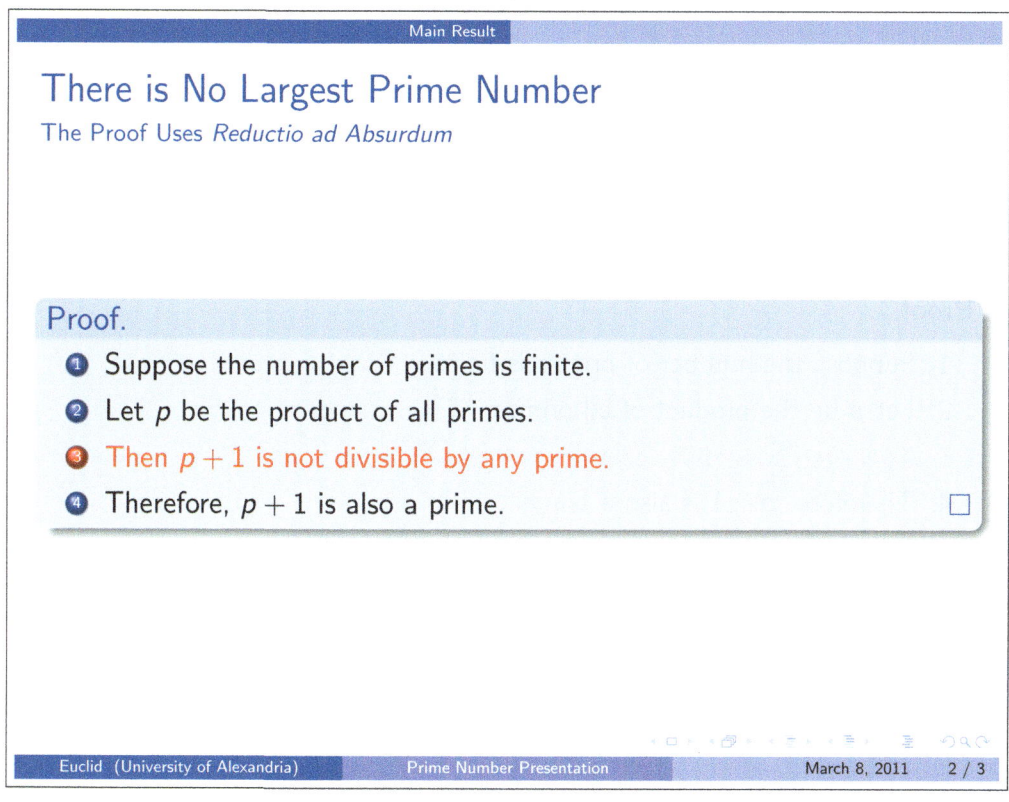

There is No Largest Prime Number
The Proof Uses *Reductio ad Absurdum*

Proof.

1. Suppose the number of primes is finite.
2. Let p be the product of all primes.
3. Then $p + 1$ is not divisible by any prime.
4. Therefore, $p + 1$ is also a prime.

Figure 14.8
Sample output of beamer's default theme. The outline of the slide is drawn for clarity.

Main Result

There is No Largest Prime Number
The Proof Uses *Reductio ad Absurdum*

Proof.

1. Suppose the number of primes is finite.
2. Let p be the product of all primes.
3. Then $p + 1$ is not divisible by any prime.
4. Therefore, $p + 1$ is also a prime.

Euclid (University of Alexandria)   Prime Number Presentation   March 8, 2011   2 / 3

Figure 14.9
Sample output of beamer's Boadilla theme. The option secheader was passed as an option to the \usetheme command. The outline of the slide is drawn for clarity.

Figure 14.10
Sample output of beamer's Antibes theme. The outline of the slide is drawn for clarity.

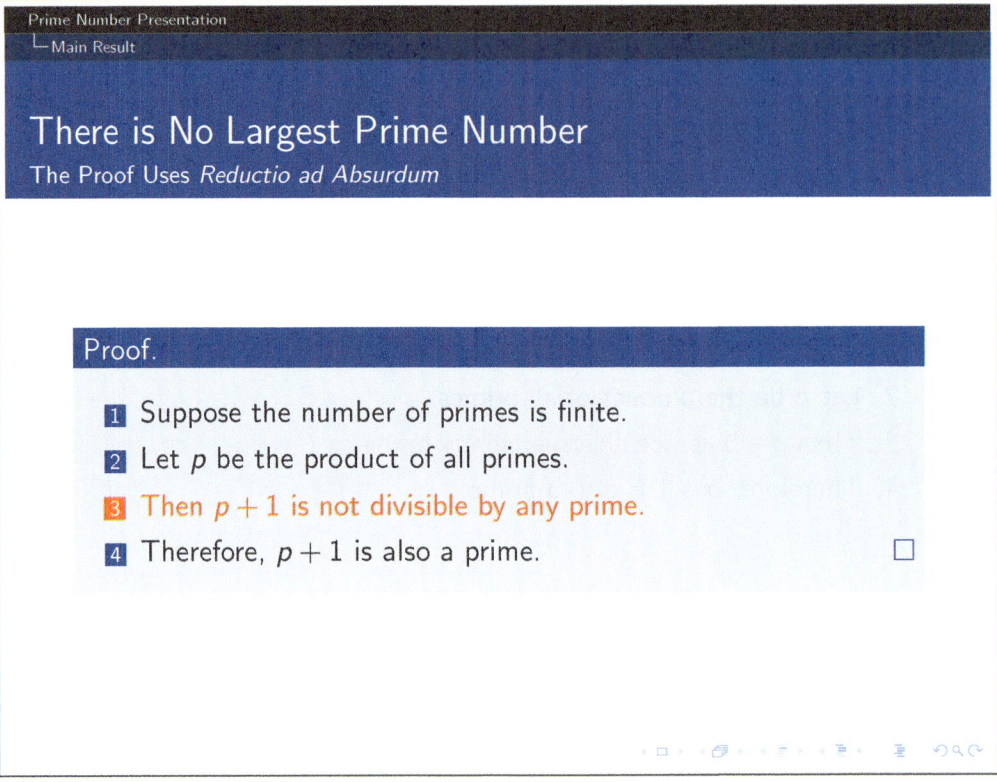

Figure 14.11
Sample output of beamer's Goettingen theme. The side bar of this theme provides more information about the structure of the presentation than the three previous themes because it also lists the names of other top-level sectional units—in this case *Conclusion*. The outline of the slide is drawn for clarity.

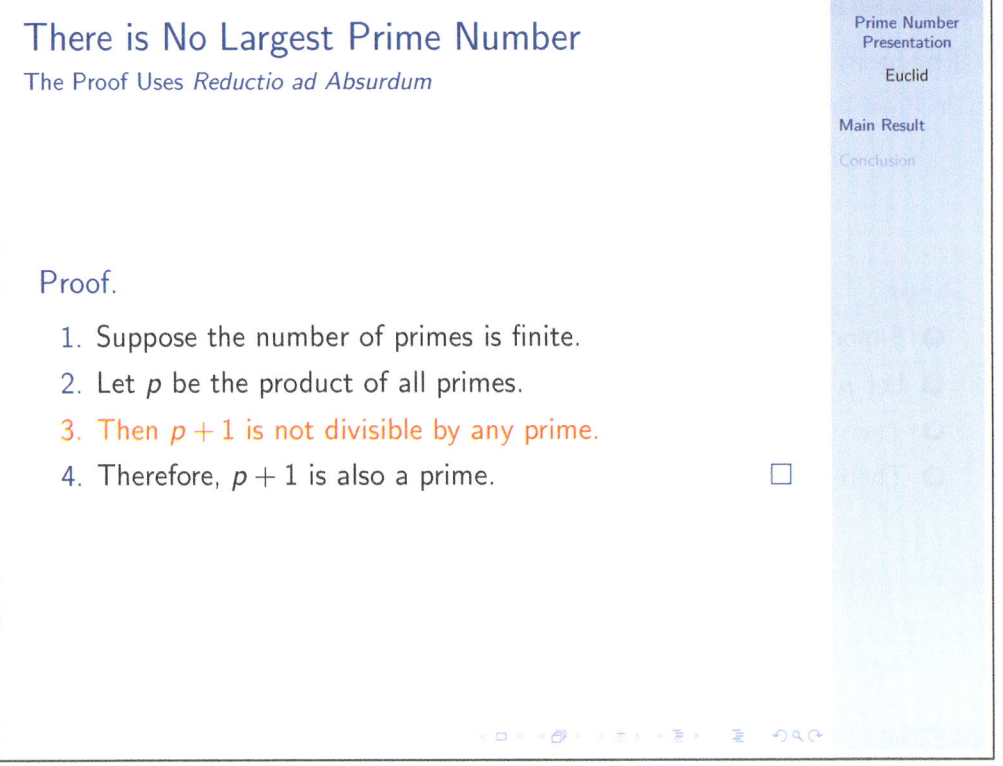

```
\documentclass{beamer}
\usetheme[⟨options⟩]{⟨theme⟩}
\usepackage{amsmath}

\title{Prime Number Presentation}
\institute{University of Alexandria}
\author{Euclid}

\begin{document}
\begin{frame}[fragile]
\maketitle
\end{frame}

\section{Main Result}

\begin{frame}[fragile]
\frametitle{There is No Largest Prime Number}
\framesubtitle{The Proof Uses \emph{Reductio ad Absurdum}}
\begin{proof}
\begin{enumerate}
\item<alert@2> Suppose the number of primes is finite.
\item<alert@3> Let $p$ be the product of all primes.
\item<alert@4> Then $p + 1$ is not divisible by any prime.
\item<alert@5> Therefore, $p + 1$ is also a prime.
                \qedhere
\end{enumerate}
\end{proof}
\end{frame}

\section{Conclusion}

\begin{frame}[fragile]
The end.
\end{frame}
\end{document}
```

Figure 14.12

Using a beamer theme. The LaTeX input is a template that is used to demonstrate the effect of the beamer themes on the previous pages. The outputs are obtained by substituting the name of the themes for ⟨theme⟩ in the input.

CHAPTER 15
Writing Classes and Packages

THIS SHORT CHAPTER is about the essence of implementing user-defined classes and packages in LaTeX2ε, which is a more recent implementation of LaTeX than Lamport's implementation.

By the end of this chapter you should know enough to write robust classes and packages that take options, parse these options, and use them to add new features on top of existing classes or packages. More complete information may be found in [LaTeX3 Project 1999]. Flynn [2007] also discusses class and package writing.

The remainder of this chapter is example driven. Besides studying some of the basics of class and package writing, we shall implement a user-defined class called modal for creating lecture presentations and lecture notes. The class is built on top of the article and beamer classes. If the beamer option is used then the class loads the beamer class but if the article option is used then the class loads the article class with a 12 pt point size. Loading the article class with the 12 pt point size shows that we can do some additional configuration. All other options are forwarded to the beamer or article class.

This chapter could also have been presented in Part V but by presenting it here we may assume enough knowledge of the beamer class, which is more convenient for the presentation.

15.1 The Structure of Classes and Packages

Class and package files are not the same as LaTeX source files. For example, class and package files are loaded in the preamble so they may not produce any output.

The main purpose of classes and packages is to set up some style-defining parameters and to provide some useful commands. They start by identifying themselves. Next they declare their options and process these options. They continue by loading auxiliary class and package files. Finally, they carry out additional configurations. This may involve installing fonts, (re)defining style parameters and commands, and so on. The remaining sections examine all steps in more detail.

15.2 Dependencies

The first thing a class or package does is state their dependencies. This is done with the following command.

`\NeedsTeXFormat{LaTeX2e}[⟨date⟩]`

> This states the dependencies of the class or package: the LaTeX implementation should be LaTeX2ε and the release date should be ⟨date⟩ or later. If you need a different LaTeX implementation, then you substitute it for `LaTeX2e`. The date should have the form ⟨four digit year⟩/⟨two digit month⟩/⟨two digit day⟩. ☑

15.3 Identification

The next thing a class or package does is identify themselves. They use the following two commands to identify themselves.

`\ProvidesClass{⟨name⟩}[⟨date⟩⟨other information⟩]`

> This formally identifies the class with the name ⟨name⟩. It is strongly recommended that you provide the optional argument. The optional argument should start with a date. The date ⟨date⟩ has the same format as explained before. The information provided in the optional argument is printed when the class is loaded. ☑

`\ProvidesPackage{⟨name⟩}[⟨date⟩⟨other information⟩]`

> This is the package version of `\ProvidesClass`. ☑

The following are the first two lines of our modal class. We start by stating the dependencies and continue by identifying the class.

```
\NeedsTeXFormat{LaTeX2e}[2009/09/24]                    LaTeX Usage
\ProvidesClass{modal}[2011/08/15 Modal class]
```

15.4 Defining and Parsing the Options

Most classes and packages take options and declaring them in LaTeX2ε is straightforward. Unfortunately, there is no standard mechanism for declaring ⟨key⟩=⟨value⟩ options. This is why we shall use the `pgfopts` package [Wright 2010], which makes `pgfkeys`-style option parsing available in classes and packages. Oberdiek [2010] also provides a package for parsing ⟨key⟩=⟨value⟩ options in classes and packages but it is not as flexible and robust as `pgfopts`. The interested reader is referred to [LaTeX3 Project 1999] for details about parsing options with the standard LaTeX2ε mechanism.

It is important to note that keys cannot contain spaces in classes and packages because they're removed by the LaTeX kernel [Wright 2010]. The following two commands trigger the option parsing.

`\ProcessPgfPackageOptions{⟨base key⟩}`

> This command triggers the option parsing for keys relative to the base key ⟨base key⟩. Any options passed to the class or package will be passed to `\pgfkeys`, which does the actual parsing. ☑

`\ProcessPgfPackageOptions*`

> This is equivalent to `\ProcessPgfPackageOptions{⟨name⟩}`, where ⟨name⟩ is the name of the current class or package. ☑

Figure 15.1 continues the implementation of our modal class. We start by declaring a TeX switch. The purpose of the switch is to deter-

```
\makeatletter
\newif\ifmodal@beamer

\pgfkeys{/modal/.cd, % definitions are relative to /modal.
        article/.style={beamer=false},
        beamer/.is if=modal@beamer,
        beamer, % turn the switch \ifmodal@beamer on
        .unknown/.code=\modal@append@option}

\newcommand*\modal@options{}
\newcommand*\modal@append@option{%
    \edef\modal@tmp{\modal@options}%
    \edef\modal@options{\modal@tmp,\pgfkeyscurrentname}%
}

% process the package options.
\ProcessPgfPackageOptions{/modal}
```

Figure 15.1

Declaring class options

mine the mode of our class: it is true if and only if the mode of the package is for a beamer presentation.

The implementation continues by calling \pgfkeys. The call starts by defining a style called /modal/article and a decision key called /modal/beamer that controls the value of the TEX switch. The style turns the switch off by executing /modal/beamer=false.

The key /modal/.unknown collects any unknown options, which are used further on. The key does this by appending any unknown option to a command called \modal@options that stores a comma-separated list of unknown options. The interested reader is invited to read the pgf manual for further information about the .unknown feature.

The call to \PgfProcessPackageOptions triggers the actual option parsing. Note that we pass the base key /modal because the pgfkeys keys are all relative to that base key.

15.5 Loading Existing Classes and Packages

Needless to say, most classes and packages are not implemented from scratch, so loading other classes and packages is usually required. The advantage of this approach is that you may implement a user-defined class file on top of another class and reuse the implementation effort. For example, the user-defined class may load the article class with all your favourite settings. Furthermore, the user-defined class may load all your favourite packages. Writing your articles with the user-defined class file is much easier because there is no more need to configure the article class. This is much easier than starting a document that uses the article class and then configuring the document.

A class file may only load one class but as many packages as it likes. A package file may also load as many packages as it likes but it may not

Figure 15.2

Loading auxiliary classes and packages

```
\ifmodal@beamer
    \LoadClass[\modal@options]{beamer}[2010/06/21]
\else
    \newif\ifmodal@contains % the variable is false now.
    \def\modal@test@containment#1[#2]{%
        \in@{#2}{#1}
        \ifin@\modal@containstrue\fi
    }
    \expandafter\modal@test@containment\modal@options[10pt]
    \expandafter\modal@test@containment\modal@options[11pt]
    \expandafter\modal@test@containment\modal@options[12pt]

    \ifmodal@contains
        \LoadClass[\modal@options]{article}[2007/10/19]
    \else
        \LoadClass[12pt,\modal@options]{article}[2007/10/19]
    \fi
    \RequirePackage{beamerarticle}[2010/05/01]
\fi

\makeatother
```

load classes. The following are the preferred commands for loading classes and packages inside class and package files.

\LoadClass[⟨options⟩]{⟨class⟩}[⟨date⟩]

This loads the class ⟨class⟩ with options ⟨options⟩ as well as any options that have been forwarded by the command \PassOptionsTo-Class. The command \PassOptionsToClass is not studied in this book but is explained in [LaTeX3 Project 1999]. The first and last argument of the \LoadClass command are optional. If you provide the last option then loading the class only succeeds if your version of the class ⟨class⟩ is at least as recent as ⟨date⟩. If the implementation of the class is older then you get an error.   ☑

\RequirePackage[⟨options⟩]{⟨package⟩}[⟨date⟩]

This loads the package ⟨package⟩. Loading a package works the same as loading a class.   ☑

\LoadClassWithOptions{⟨class⟩}[⟨date⟩]

This loads ⟨class⟩ with the same options as the current class.   ☑

\RequirePackageWithOptions{⟨package⟩}[⟨date⟩]

This loads ⟨package⟩ with the same options as the current package. ☑

15.6 Final Configuration

At this stage we only have to do some extra configuring and load the auxiliary classes and packages. This is shown in Figure 15.2.

Let's examine the code in Figure 15.2 in more detail. The switch \ifmodal@beamer indicates the mode of our class. We load the beamer class if the switch is on; otherwise we load the article class.

Loading the beamer class is easy. When we load it we also pass the unknown options to the class. There is no need to expand the unknown options.

Loading the article class is a bit more tricky. The reason is that we want to use the class with a default point size of 12 pt. We cannot simply add the 12pt option to the unknown options because the unknown options may already contain a valid point size option. Remember that the article class has only three valid point size options: 10pt, 11pt, and 12pt.

There are two obvious approaches to determining whether there is a valid point size in the unknown options. The first approach is to use the \pgfkeys command and parse the options. This is really easy to do and we already know the technique.

The second approach is to do the parsing ourselves. This is more difficult and we will learn something new, so lets use this approach. We start by introducing a new switch called \ifmodal@contains. When this switch is created it is off by default. For each valid point size 10pt, 11pt, and 12pt we turn the switch on if it is contained in the unknown options.

To implement our own parsing we shall use the macro \in@, which is provided by the LaTeX kernel [Scharrer 2011-07-23]. The call \in@{⟨elem⟩}{⟨list⟩} tests if ⟨elem⟩ is in the comma-delimited list ⟨list⟩. The macro sets the switch \ifin@ accordingly. Unfortunately both arguments of \in@ must be expanded, so we must do a bit more work to expand the unknown options. This is what the TeX delimited macro \modal@contains and the three lines after it are all about.

The rest is straightforward. If the switch \ifmodal@contains is off we load the article with the 12pt option; otherwise we load it without the 12pt option. We conclude by loading the auxiliary package beamerarticle. This package is used in combination with the article class. Among others it makes sure that all beamer commands work in article mode.

Congratulations. You've just implemented a full blown LaTeX class. Who would have thought it would be that easy.

CHAPTER 16
Using OpenType Fonts

THE OPENTYPE FONT FORMAT is an extension of the TrueType font format. The format was developed by Microsoft and Adobe and supports different glyph variants. For example, a font may have glyph variants for a zero with and without a slash. A single OpenType font may support different variants, which are organised by features.

This relatively technical chapter explains how to use OpenType fonts with `pdflatex`. As a matter of fact it explains how to implement the font-related commands that were used to typeset this book.

Using OpenType fonts is really easy with Will Robertson and Khaled Hosney's `fontspec` package [Robertson 2011]. The package gives you access to all OpenType glyphs, which may be selected by their OpenType features. Unfortunately, `fontspec` requires X$_{\overline{3}}$TEX or LuaTEX and doesn't work with pdfLATEX.

OpenType fonts typically have several thousands of glyphs. LATEX fonts are limited to 256 glyphs, which rules out including a single OpenType font as a single LATEX font. To overcome this problem we shall extract the glyphs from the OpenType font and embed them into a series of LATEX fonts. We shall add some extra structure by letting each LATEX font consist of glyphs that agree on a set of carefully chosen OpenType features. For example, one of our LATEX fonts has proportional oldstyle numbers such as 0123456789, another font has non-proportional (tabular) lining numbers such as 0123456789, and so on. We shall define commands for switching from one feature set to another. For example, if the current mode uses proportional oldstyle numbers, then a switch to tabular numbers will result in tabular oldstyle numbers.

In the remainder of this chapter we shall use a commercial font called *Nexus* as an example. This is the main font that was used to typeset this book. To illustrate the possibilities, we shall install fonts for four different configurations of numbers: proportional oldstyle numbers, tabular oldstyle numbers, proportional lining numbers, and tabular lining numbers. Also we shall install the commands that let you access these fonts. It should be possible, at least in principle, to automate this chapter's approach. However, the biggest limitation is that the number of required files is exponential in the number of features that can be combined.

Throughout this chapter it is assumed that you know how to im-

Table 16.1

Some font features of the italic, seriffed shape of the Open-Type font *Nexus*. The columns 'Feature' list the 4-letter feature abbreviation. The columns 'Description' explain the meaning of the abbreviation. Strictly speaking onum isn't available. However, this is the default figure feature.

Feature	Description	Feature	Description
aalt	Access All Alternates	ornm	Ornaments
c2sc	Small Capitals From Capitals	pnum	Proportional Numbers
cpsp	Capital Spacing	smcp	Small Capitals
dnom	Denominators	ss01	Stylistic Set 1
liga	Standard Ligatures	...	...
lnum	Lining Numbers	ss06	Stylistic Set 6
numr	Numerators	subs	Subscript
onum	Oldstyle Figures	sups	Superscript
ordn	Ordinals	tnum	Tabular Figures

plement LaTeX and TeX commands and that you know the pgfkeys option parsing techniques that are studied in Chapter 13.

16.1 OpenType Font Features

In this example-driven section we shall use Eddie Kohler's otfinfo command to study the features of OpenType fonts. If you supply the option --features and the name of the OpenType font file, the command will output a list consisting of the font's font features. The following shows how this works.

```
$ otfinfo --features NexusSerifOT-Italic.otf          Unix Session
aalt    Access All Alternates
c2sc    Small Capitals From Capitals
...
```

The output of the previous example is limited to a few lines for brevity. Table 16.1 lists some of the more useful font features. The remainder of this section briefly explains some of these features.

c2sc/smcp The properties c2sc and smcp are related to small caps letters. Many LaTeX fonts implement "faked small caps" glyphs by scaling uppercase letters. Good fonts have special glyphs for small caps numbers. These glyphs are sometimes referred to as "real small caps." The features smcp and c2sc stand for small caps and capitals to small caps. The feature smcp turn lowercase letters into real small caps, whereas c2sc turns uppercase letters into real small caps. If you can, you should use real small caps glyphs because they usually look better.

ornm Some fonts have special ornament glyphs, which are intended for decoration. Common ornaments are fleurons but ornament glyphs are not standardised. For example *Nexus* has "ornaments" like ⊕, ⊠, ⟨, ©, ⟨, and so on. Glyphs like this have the property ornm.

ss01–ss06 Some fonts have stylistic alternates for some of the characters. A beautiful example is the *Zapfino* font, which has nine stylistic alternates for some of the characters. Figure 16.1 depicts an example. *Nexus also has stylistic alternate glyphs.* Accessing stylistic alternate glyphs is not explained here but it is not difficult to implement.

Figure 16.1
Stylistic alternates. This example shows eight stylistic alternates of the the letter *d* in the *Zapfino* font. This example is included with the kind permission of Dario Taraborelli.

	Oldstyle	Lining
Proportional	01111888823456789	01111888823456789
Proportional	08888111123456789	08888111123456789
Tabular	08888111123456789	08888111123456789
Tabular	01111888823456789	01111888823456789

Table 16.2
Figure feature combinations. This table lists four combinations of figure feature that are available in many modern OpenType fonts. The rows list the width-affecting features. The columns list the height-affecting features.

onum/lnum There are several properties that are related to figures (digits). The properties onum and lnum affect the height of the figure glyphs. The property onum is for oldstyle numbers, whereas the property lnum is for lining numbers. The height of oldstyle numbers varies, whereas the heights of lining numbers are all the same. Oldstyle numbers look better in the running text. For example, the number 1234567890 is in oldstyle numbers, which blend in well with the text but 1234567890 is in lining numbers, which look too high for the main text. Lining numbers look better in text with uppercase letters: MAIN PRIZE: BANK OF SANS SERIFFE REWARD CHECK TO THE VALUE OF $ 1,000,000 (HEX). Remember that you should always letterspace texts like this. The easiest way to letterspace text is with the \textls command, which is provided by the microtype package.

pnum/tnum The features pnum and tnum affect the width of the figures. Glyphs with the property tnum (tabular number) have a fixed width. Tabular numbers are ideal for tables with numeric data because the figures in the numbers align nicely in columns. Glyphs with the property pnum (proportional number) have proportional width. These look better in the running text. Table 16.2 depicts four different combinations of figure features.

dnom/numr Some fonts have special glyphs for frequently occurring fractions. They may also provide special glyphs for figures in numerators and denominators of fractions. These glyph variants have the property numr for numerators and the property dnom for denominators.

To see why these glyph variants are interesting, consider the fraction 42/133, which uses oldstyle numbers and an ordinary slash. Such fractions will never win the first prize in a beauty competition. Substituting lining figures for the oldstyle numbers improves things a little bit: 42/133. However, $^{42}\!/_{133}$ was produced with a special command that uses a solidus (virgule) for the slash and the proper glyphs for the numerator and denominator. The kerning of the numerator and denominator is currently handled by the command. For example, if the numerator ends in a 7 then the solidus should be moved a bit further to the left than if the numerator ends in a 1. The kerning for the denominator uses differen rules. For example, if the denominator starts with a 4 then it should be moved further to the left than if the numerator starts with a 1. This kind of ad hoc kerning can also be handled at the font level.

The idea to create the special command for fractions is inspired by Michael Saunders. Reichert [1998] and Høgholm [2011] propose different solutions for typesetting fractions. With Reichert's nicefrac package you get fractions like 33/71 and 11/42 for oldstyle numbers and fractions like 33/71 and 11/42 for lining numbers. Høgholm's xfrac package uses a solidus by default. With this approach you get fractions like $^{33}\!/_{71}$, $^{11}\!/_{42}$, $^{33}\!/_{71}$, and $^{11}\!/_{42}$. The reader is invited to compare these fractions and the fractions $^{33}\!/_{71}$ and $^{11}\!/_{42}$, which are typeset with the solidus and the glyph variants for numerators and denominators.

subs/sups The features subs and sups are glyph variants for subscripts and superscripts. For example, consider 'text,^{1},' which you get if you use LaTeX's standard superscript construct 'text,^{1},' which scales the figure 1. Next consider 'text,^{1}' which you get if you use the special glyph of the figure 1 for superscripts. The special glyph is much larger than the scaled 1 and is easier to read. Note that it may look better if we add a bit of kerning to force the 1 on top of the comma: text^1.

16.2 LaTeX Font Selection Mechanism

This section is a brief introduction to LaTeX's font selection mechanism It is mainly based on [LaTeX3 Project 2000], [Mittelbach, Fairbairns, and Lemberg 2006], [Rahts 1993], and [Goossens, Rahtz, and Mittelbach 1997].

The key to LaTeX2_ε font selection is setting the right values for the right attributes. LaTeX2_ε has five font attributes, which are as follows.

encoding This attribute specifies the order of the glyphs in the font. There are several kinds of encodings but the most commonly used encodings for text are OT1 for TeX text and T1 for TeX extended text. In our implementation we shall use an encoding called LY1, which is also known as Y&Y encoding [Mittelbach, Fairbairns, and Lemberg 2006]. Encodings are defined in font encoding files. A position in an encoding is also known as a slot. By putting the right glyph in the right slot in an encoding file, you can make LATEX print the glyph for the character that is supposed to be in that slot. For example, the lowercase letters in an encoding file are stored in slots 61–7A (hexadecimal). By putting small caps glyph variants in those slots, you can make LATEX print a small caps A when you use a lowercase *a* in your input. Figure 16.2 shows this in further detail.

family The font family attribute defines a collection of related fonts. Typically, a LATEX font family provides a roman shape, an italic shape, a slanted shape, a small caps shape, and so on.

Usually, the members of a font family are from the same designer, but this is not required—especially in LATEX font families. The main criterion for a good font family is that its members get on well. Examples of font families are *Computer Modern Roman, Computer Modern Sans Serif, Gentium, Helvetica,* and so on.

LATEX font families are defined in font definition files. How to create font definition files is explained further on.

series The series attribute determines the blackness of a font. For example, the roman shape may have specially defined font files for medium weight, for bold, for bold extended, and so on. The series is characterised by values such as l for light, m for medium, b for bold, bx for bold extended, and so on. The most common series have standardised values, but you may also define your own values.

shape The shape attribute determines, well, the shape of the font: upright (roman), italic, slanted, small caps, and so on. Shapes are also characterised by letter sequences. The most common shape values have standard names: n for normal/upright, it for italic, sl for slanted, and sc for small caps. As with the series attribute you may also define your own values.

size The last font attribute is the size of the font. The size of a font is the sum of the largest height and the largest depth of the bounding boxes of the letters in the font. Some fonts have different designs for different sizes. For example, they may have one design for 6 pt, a different design for 10 pt, and so on. Many fonts only have one design at one fixed type size. This font is scaled to different type sizes. If there are designs for different type sizes then a font may be scaled from one of these designs. Usually it is scaled from the design with a similar type size. This chapter assumes there is only one design for each font. LATEX3 Project [2000] explains how to use several designs.

For each of these previous five font attributes there is a command that sets the attribute. The following are these commands.

`\fontencoding{⟨encoding⟩}`

This sets the current font encoding to ⟨encoding⟩. In this book we shall always use the font encoding LY1. ☑

`\fontfamily{⟨family⟩}`

This sets the current font family to ⟨family⟩. ☑

`\fontseries{⟨series⟩}`

This sets the current font series to ⟨series⟩. This command is called

Figure 16.2

A typical font encoding file. The file was generated by otftotfm and defines what is in the 256 slots of an LY1 encoding. The lines starting with a percentage sign are comments. Most comments list the number of the next slot in hexadecimal. Each slot specification is of the form /⟨spec⟩. Empty slots are denoted /.notdef. For the remaining slots ⟨spec⟩ is the postscript or Unicode name of the glyph in the slot. The order of the slot specifications is significant. For example, there is no glyph in slot 0. The glyph in slot 1 is the Euro glyph, the next two slots are empty, the glyph in slot 4 is the virgule/solidus, and so on. The main purpose of the encoding file in this figure is to support a LATEX font that uses small caps glyphs for the lowercase letters. This is done by putting special small caps glyphs in the slots that are reserved for the lowercase letters. For example, the LY1 encoding reserves slots 61–7A for lowercase letters. The glyphs in these slots all have names ending in .sc, which indicates that they're small caps glyph variants of these letters. Some other slots are also occupied by small caps glyphs. For example, there's a small caps exclamation mark in slot 21, a small caps question mark in slot 3F, and so on.

```
% THIS FILE WAS AUTOMATICALLY GENERATED -- DO NOT EDIT

%%AutoEnc_nm2xhbpmn36tu3xg5fo7gsmswg
% Encoding created by otftotfm on Wed Aug  3 04:14:15 2011
% Command line follows encoding
/AutoEnc_nm2xhbpmn36tu3xg5fo7gsmswg [
%00
    /.notdef /Euro /.notdef /.notdef /fraction.sc /dotaccent /hungarumlaut /ogonek
    /fl.sc /.notdef /.notdef /uniFB00 /fi.sc /.notdef /uniFB03 /uniFB04
%10
    /dotlessi.sc /.notdef /grave /acute /caron /breve /macron /ring
    /cedilla /germandbls.sc /ae.sc /oe.sc /oslash.sc /AE /OE /Oslash
%20
    /space /exclam.sc /quotedbl /numbersign.sc /dollar.sc /percent.sc /ampersand.sc /quoteright
    /parenleft.sc /parenright.sc /asterisk.sc /plus /comma /hyphen /period /slash.sc
%30
    /zero /one /two /three /four /five /six /seven
    /eight /nine /colon /semicolon /less /equal /greater /question.sc
%40
    /at /A /B /C /D /E /F /G
    /H /I /J /K /L /M /N /O
%50
    /P /Q /R /S /T /U /V /W
    /X /Y /Z /bracketleft.sc /backslash.sc /bracketright.sc /circumflex /underscore
%60
    /quoteleft /a.sc /b.sc /c.sc /d.sc /e.sc /f.sc /g.sc
    /h.sc /i.sc /j.sc /k.sc /l.sc /m.sc /n.sc /o.sc
%70
    /p.sc /q.sc /r.sc /s.sc /t.sc /u.sc /v.sc /w.sc
    /x.sc /y.sc /z.sc /braceleft /bar.sc /braceright /tilde /dieresis
%80
    /Lslash /quotesingle /quotesinglbase /florin /quotedblbase /ellipsis.sc /dagger /daggerdbl
    /.notdef /perthousand.sc /Scaron /guilsinglleft /.notdef /Zcaron /asciicircum /minus
%90
    /lslash.sc /.notdef /.notdef /quotedblleft /quotedblright /bullet /endash /emdash
    /.notdef /trademark /scaron.sc /guilsinglright /.notdef /zcaron.sc /asciitilde /Ydieresis
%A0
    /uni00A0 /exclamdown.sc /cent /sterling /currency /yen /brokenbar.sc /section
    /.notdef /copyright /ordfeminine.sc /guillemotleft /logicalnot /uni00AD /registered /.notdef
%B0
    /degree /plusminus /twosuperior /threesuperior /.notdef /mu /paragraph /periodcentered
    /.notdef /onesuperior /ordmasculine.sc /guillemotright
    /onequarter.sc /onehalf.sc /threequarters.sc /questiondown.sc
%C0
    /Agrave /Aacute /Acircumflex /Atilde /Adieresis /Aring /.notdef /Ccedilla
    /Egrave /Eacute /Ecircumflex /Edieresis /Igrave /Iacute /Icircumflex /Idieresis
%D0
    /Eth /Ntilde /Ograve /Oacute /Ocircumflex /Otilde /Odieresis /multiply
    /.notdef /Ugrave /Uacute /Ucircumflex /Udieresis /Yacute /Thorn /.notdef
%E0
    /agrave.sc /aacute.sc /acircumflex.sc /atilde.sc /adieresis.sc /aring.sc /.notdef /ccedilla.sc
    /egrave.sc /eacute.sc /ecircumflex.sc /edieresis.sc
    /igrave.sc /iacute.sc /icircumflex.sc /idieresis.sc
%F0
    /eth.sc /ntilde.sc /ograve.sc /oacute.sc /ocircumflex.sc /otilde.sc /odieresis.sc /divide
    /.notdef /ugrave.sc /uacute.sc /ucircumflex.sc /udieresis.sc /yacute.sc /thorn.sc /ydieresis.sc
] def
% Command line: 'otftotfm -v fsi -fkern -fliga --ligkern T{L}h
% --map-file=fsi.map --no-updmap --force -a -e texnansx -fonum -fpnum
% -fsmcp NexusSerifOT-Regular.otf LY1--NexusSerifOT-Regular--fonum--fpnum--fsmcp'
```

by commands such as `\mdseries` and `\texnormal`, `\bfseries` and `\textbf`, and so on. ☑

`\fontshape{⟨shape⟩}`

This sets the current font shape to ⟨shape⟩. This command is called by commands such as `\rmfamily` and `\textrm`, `\itshape` and `\textit`, and so on. ☑

`\fontsize{⟨size⟩}{⟨unspread leading⟩}`

This sets the point size of the font to ⟨size⟩ and the leading to the product of ⟨unspread leading⟩ and the current value of the *line spread*. The leading is stored in a length command that is called `\baselineskip`. The notion of line spread is explained in the next paragraph. The command `\fontsize` is called by commands such as `\scriptsize`, `\large`, and so on. ☑

In addition, there is one command that sets the line spread.

`\linespread{⟨factor⟩}`

This command defines a multiplication factor that is called the line spread. When there is a call to `\fontsize{⟨size⟩}{⟨unspread leading⟩}`, the value of `\baselineskip` is set to the product of ⟨unspread leading⟩ and the line spread. It is strongly advised to set the line spread to 1. ☑

You may use any combination of the previous six commands but a call to `\selectfont` should follow them before you typeset anything.

`\selectfont`

This loads the font that is determined by the current values of the font attributes. Changes to font attribute values may not result in a change of the type style unless they are immediately followed by a call to `\selectfont`. For example, '`\fontshape{it} \selectfont first \fontshape{sc} second`' may typeset both 'first' and 'second' in italic. However, '`\fontshape{it} \selectfont first \fontshape{sc} \selectFont second`' should typeset 'first' in italic and 'second' in small caps. ☑

16.3 Overview of Functionality

This section briefly explains the functionality we are going to implement in the remainder of this chapter. Basically, we want to implement commands that give access to some useful glyphs in an existing Open-Type font with the same ease as the commands that are provided by the `fontspec` package. As an example of what is possible, we shall implement some user-defined commands that give access to different glyphs for figures (digits).

With the `fontspec` package you can access glyph variants by their OpenType features. For example after the following commands the roman text is typeset in the font `NexusSerifOT-Regular` with oldstyle proportional numbers.

```
\usepackage{fontspec}
\setromanfont[Numbers={OldStyle,Proportional}]
          {NexusSerifOT-Regular}
```

LATEX Usage

The previous commands make sure that the roman text is typeset using the external OpenType font NexusSerifOT-Regular and that the glyph variant combination onum/pnum is used for numbers. However, you still have access to different number glyph variants.

```
1234567890 and
\fontspec[numbers={Lining}]        1234567890 and 1234567890.
        {1234567890.}
```

This chapter implements a similar functionality but with different commands. For example, we want declarations that enable (switch to) a given font feature and commands that take an argument and typeset the argument according to the command's feature. For example,

```
\TLNumSwitch 1234567890
\ONumStyle{and 1234567890.}        1234567890 and 1234567890.
\PNumSwitch 1234567890             1234567890 and 1234567890.
\ONumStyle{and 1234567890}.
```

In this example, the command \TLNumSwitch is a declaration that switches to tabular lining numbers. The command \ONumStyle sets the numbers in its argument in oldstyle number format. When the command \ONumStyle is called, it will maintain the tabular feature because it is compatible with oldstyle numbers. The command \PNum-Switch is a declaration that switches to proportional number style. Note that the argument of the command \ONumStyle may also contain non-numeric characters.

16.4 Inspecting the Font

One of the first things you want to do is to see the glyphs of your OpenType font. Again Eddie Kohler comes to the rescue with his cfftotl and tltestpage tools. The following unix script, which is called TestPage, lets you extract the glyphs from a list of OpenType font files:

```
#!/bin/sh                                              Unix Usage


for FILE in $@; do
    BASE=`basename ${FILE} .otf`
    cfftotl ${FILE} | tltestpage |
            epstopdf --filter > ${BASE}.pdf
done
```

Running the script on the *Nexus* OpenType file that is called NexusSerifOT-Italic.otf is done as follows.

```
$ ./TestPage NexusSerifOT-Italic.otf                   Unix Usage
```

Figure 16.3 depicts the first page of the resulting output. The figure shows the available glyphs for the first letters. Notice that each glyph has a name.

NexusSerifOT-Italic

A	A.ss01	A.ss02	a	a.dnom	a.numr	a.sc	a.ss03	a.ss04	a.subs
a.sups	B	B.ss01	B.ss02	b	b.dnom	b.numr	b.sc	b.ss03	b.ss04
b.subs	b.sups	C	C.ss01	C.ss02	c	c.sc	c.ss03	c.ss04	D
D.ss01	D.ss02	d	d.dnom	d.numr	d.sc	d.ss03	d.ss04	d.subs	d.sups
E	E.ss01	E.ss02	e	e.dnom	e.numr	e.sc	e.ss03	e.ss04	e.subs
e.sups	F	F.ss01	F.ss02	f	f.sc	f.ss03	f.ss04	G	G.ss01
G.ss02	g	g.sc	g.ss03	g.ss04	H	H.ss01	H.ss02	h	h.dnom
h.numr	h.sc	h.ss03	h.ss04	h.subs	h.sups	I	I.ss01	I.ss02	i
i.dnom	i.latn_TRK.sc	i.numr	i.sc	i.ss03	i.ss04	i.subs	J	J.ss01	J.ss02
j	j.sc	j.ss03	j.ss04	K	K.ss01	K.ss02	k	k.sc	k.ss03
k.ss04	L	L.ss01	L.ss02	l	l.dnom	l.numr	l.sc	l.ss03	l.ss04
l.subs	l.sups	M	M.ss01	M.ss02	m	m.dnom	m.numr	m.sc	m.ss03
m.ss04	m.subs	m.sups	N	N.ss01	N.ss02	n	n.dnom	n.numr	n.sc

Figure 16.3

Sample of some of the glyphs of one of the *Nexus* variants. The output was created with the aid of Eddie Kohler's cfftot1 and t1testpage tools. The output is scaled to 70 % of the original size.

It is also possible to inspect the names of the glyphs with `otfinfo`. The following shows how this is done.

```
$ otfinfo --glyphs NexusSerifOT-Italic.otf          Unix Usage
# 787 lines with glyph names omitted.
```

The output is a long list of lines, each consisting of the name of a single glyph. The following are the names for the glyphs of the uppercase and the lowercase *a:* A (uppercase *A*), A.ss01 (stylistic alternate 1), A.ss02 (stylistic alternate 2), a (lowercase *a*), a.dnom (denominator), a.numr (numerator), a.sc (small caps), a.ss03 (stylistic alternate 3), a.ss04 (stylistic alternate 4), a.subs (subscript), and a.sups (superscript). It is reassuring that these names are the same as the ones listed at the top of Figure 16.3.

16.5 Current Alternatives

There are some other approaches to using OpenType fonts. Owens [2006] provides tools to generate a package from an OpenType font specification. Basically the tools are a wrapper around `otftotfm`. The user supplies a so-called *Berry* code [Berry 1990] and scaling information. Choosing the Berry code requires some technical knowledge from the user. If the Berry code is not known, the user must edit a script. The script generates the required LaTeX font files and a package that can be loaded in oldstyle mode or in lining mode. The package does not provide support to select other font features. If all you need is an OpenType font for one single feature (oldstyle/lining) then the solution presented by Owens is easier than the solution proposed in this chapter.

As explained in the introduction, the `fontspec` package also provides support for OpenType fonts. However, the package does not support `pdftex`.

16.6 Designing the Font Families

In this section we shall design the font families. *Nexus* has four families: serif, sans serif, mix, and typewriter. The mix variants is a slab serif: the quick brown fox jumps over the lazy dog. For simplicity we shall ignore the mix variant. The remaining variants come with a regular and italic shape and a normal and bold weight (series). The seriffed and sans serif variants have oldstyle tabular (ot) numbers, lining tabular (lt) numbers, oldstyle proportional (op) numbers, and lining proportional (lp) numbers. Not entirely surprising the typewriter variants only have ot and lt numbers. The seriffed and sans serif variants also have special glyphs for small caps for each combination of font shape and font weight. Table 16.3 lists all possible combinations. The total number of combinations is $2 \times 2 \times 2 \times 4 \times 2 + 1 \times 2 \times 2 \times 2 \times 1 = 64 + 8 = 72$ combinations of font variant, shape, weight, and number features.

To implement our font switching commands we shall use 12 font font families: one font family for each combination of font shape (serif,

Font families:	Serif	Sans Serif	Typewriter
font shape			
regular	√	√	√
italic	√	√	√
font weight			
normal	√	√	√
bold	√	√	√
number features			
ot	√	√	√
lt	√	√	√
op	√	√	
lp	√	√	
small caps			
disabled	√	√	√
enabled	√	√	

Table 16.3

Nexus font features. This table list all possible combinations of font family, font shape, font weight, and number and small caps features. The columns are the font families. A √ in the row for a given shape, weight, or feature means that the corresponding variant has the shape, weight, or feature.

sans serif, and typewriter) and number features. Two of these families are artificial because the typewriter shape doesn't have proportional numbers. However, introducing them simplifies the overall design.

By cleverly naming the 12 font families we can determine the font features from the name of the families. Given the font features we can then switch to a different font family if we want to change one of the features of the current font family. In the following we shall use font family names of the form ⟨font variant⟩--⟨fonum or flnum⟩--⟨ftnum or fpnum⟩. Lets assume that serif--fonum--ftnum is our current font family then this means than tabular oldstyle numbers are active. If we wanted to use lining numbers then we simply change the feature onum to lnum: serif--flnum--ftnum. With TeX's delimited macros such changes are easy to implement. For example, let's assume we have the following definition.

```
\def\change@first@feature#1--f#2--#3[#4]{%      LaTeX Usage
    #1--#4--#3%
}
```

Then \change@first@feature NexusSerif--fonum--ftnum[lnum] gives us NexusSerif--flnum--ftnum. Changing the font variant and changing other features can be implemented in a similar way. Different techniques are also possible.

16.7 Extracting the Fonts

Creating and installing LaTeX fonts from scratch may be difficult and time consuming. This is partially because of decisions about naming conventions and file locations. Eddie Kohler's `otftotfm` [Kohler 2011] has made the installation process of OpenType fonts extremely easy.

Given an OpenType font, you tell `otftotfm` about the features you are interested in and it will create the font encoding (`.enc`) files, the font metric (`.tfm`) files, and the virtual font (`.vf`) files, and put them in the right place. We have already seen encoding files: they determine the order of the characters in the fonts. Font metric files tell LaTeX about the sizes of the characters, kerning, and rules for forming from character combinations. The combination of a `tfm` and a `vf` file defines a virtual font [Goossens, Rahtz, and Mittelbach 1997, Chapter 10]. For our purpose the `vf` file is used to find the glyphs in the OpenType font. In the remainder we shall treat the `tfm` and `vf` files as if they were the same. We shall refer to them as font metric files. From a conceptual point of view this suffices to know how to *use* them—not how they work.

In the rest of this section we shall use `otftotfm` to create the font encoding and font metric files for one of the members of one font family. Creating the remaining font encoding and font metric files works in a similar way.

The `otftotfm` uses lots of command line arguments. In the following we define a variable that defines the default flags.

```
FLAGS="-v fsi -fkern -fliga --ligkern T{L}h \            Unix Usage
        --no-updmap --map-file=fsi.map \
        --force -a -e texnansx"
```

Most of these flags are recommended. The flags `--ligkern T{L}h` tell `otftotfm` not to turn the combination `Th` into a ligature. The flags `-v fsi` tell `otftotfm` the vendor is `fsi`—a shorthand for *Font Shop International*. The flag `--no-updmap` tells `otftotfm` not to update the default font map files and the TeX name database. Using the flag makes `otftotfm` significantly faster. However, if you do use it, you have to explicitly update the name database afterwards. Note that the flag `-fliga` turns standard ligature combinations. You probably want to turn this flag off when the fonts for the typewriter family.

The following shows how to create the font encoding and font metric files for the number features `onum` and `pnum`. We create files for the regular shape with and without small caps. Our encoding is `LY1`, which is why the names of the font metric files (the last command line arguments) start with `LY1--`.

```
$ otftotfm ${FLAGS} \                                    Unix Usage
        NexusSerifOTRegular.otf -fonum -fpnum \
        LY1--NexusSerifOTRegular--fonum--fpnum
$ otftotfm ${FLAGS} \
        NexusSerifOTRegular.otf -fonum -fpnum -fsmcp \
        LY1--NexusSerifOTRegular--fonum--fpnum--fsmcp
```

Creating the remaining files is done in a similar way but for the italic fonts you have to provide the italic angle because the font files don't provide information about the italic angle. Adding the flag `--`

```
\ProvidesFile{NexusSerif--flnum--fpnum.fd}
              [2011/07/26 v1.0 Nexus Support]


\DeclareFontFamily{LY1}{NexusSerif--flnum--fpnum}{}
\DeclareFontShape{LY1}{NexusSerif--flnum--fpnum}{m}{n}
     {<-> LY1--NexusSerifOT-Regular--flnum--fpnum}{}
...
\DeclareFontShape{LY1}{NexusSerif--flnum--fpnum}{b}{it}
     {<-> LY1--NexusSerifOT-BdIt--flnum--fpnum}{}
...
\endinput
```

Figure 16.4
Partial font definition file

`italic-angle=-7` to `otftotfm` should do the trick. Normally, there's no need to supply the italic angle.

When you're finished creating the font files, you have to move the map file to its proper destination. This is done as follows.

```
$ mkdir -p ${TEXMFVAR}/fonts/map/dvips/fsi
$ mv fsi.map ${TEXMFVAR}/fonts/map/dvips/fsi
```
Unix Usage

Finally, we update the default font map files and update the TeX filename database.

```
$ updmap --enable Map fsi.map
```
Unix Usage

16.8 Font Definition Files

Remember that a font family is a collection of related fonts. LaTeX defines each font family in its own font definition file. These files come with the extension `fd`. The purpose of the font definition file is to relate the five previously mentioned attributes with the name of the font family. For example, the font definition file may define a font family that uses uses font ⟨roman⟩ at ⟨size⟩ for the regular shape and normal series, uses font ⟨roman bold⟩ at ⟨size⟩ for the regular shape and bold series, uses font ⟨italic⟩ at ⟨size⟩ for the italic shape and normal series, and so on.

Figure 16.4 is an example of a font definition file that defines a font family called `NexusSerif--flnum--fpnum.fd` for the `LY1` font encoding. The base name of the font definition file is obtained by adding the name of the font encoding before the name of the font family. In this case the name is `LY1NexusSerif--flnum--fpnum.fd`. In the font definition file, the name `LY1--NexusSerifOT-Regular--flnum--fpnum` and the name `LY1--NexusSerifOT-BdIt--flnum--fpnum` are names of font metric files, which were generated in the previous section. The following explains the commands in Figure 16.4 in further detail.

`\DeclareFontFamily{⟨enc⟩}{⟨family⟩}{⟨commands⟩}`
This declares the name of the font family. The argument ⟨enc⟩ is the font encoding and ⟨family⟩ is the name of the font family. The

Figure 16.5
Nexus font definition file

```
\ProvidesFile{NexusSerif--fonum--fpnum.fd}
             [2011/07/26 v1.0 Nexus Support]

\DeclareFontFamily{LY1}{NexusSerif--fonum--fpnum}{}
\DeclareFontShape{LY1}{NexusSerif--fonum--fpnum}{m}{n}
     {<-> LY1--NexusSerifOT-Reg--fonum--fpnum}{}
\DeclareFontShape{LY1}{NexusSerif--fonum--fpnum}{m}{sl}
     {<-> LY1--NexusSerifOT-It--fonum--fpnum}{}
\DeclareFontShape{LY1}{NexusSerif--fonum--fpnum}{m}{it}
     {<-> LY1--NexusSerifOT-It--fonum--fpnum}{}
\DeclareFontShape{LY1}{NexusSerif--fonum--fpnum}{b}{n}
     {<-> LY1--NexusSerifOT-Bd--fonum--fpnum}{}
\DeclareFontShape{LY1}{NexusSerif--fonum--fpnum}{b}{sl}
     {<-> LY1--NexusSerifOT-BdIt--fonum--fpnum}{}
\DeclareFontShape{LY1}{NexusSerif--fonum--fpnum}{b}{it}
     {<-> LY1--NexusSerifOT-BdIt--fonum--fpnum}{}

\DeclareFontShape{LY1}{NexusSerif--fonum--fpnum}{m}{sc}
     {<-> LY1--NexusSerifOT-Reg--fonum--fpnum--fsmcp}{}
...
\endinput
```

⟨commands⟩ are executed when the font family member is loaded. In this chapter we won't use such commands. ☑

\DeclareFontShape{⟨enc⟩}{⟨family⟩}{⟨series⟩}{⟨shape⟩}{⟨info⟩}{⟨commands⟩}
This declares a member of the font family ⟨family⟩ for font encoding ⟨enc⟩. The arguments ⟨series⟩ and ⟨shape⟩ define the series and shape of the member. The example in Figure 16.4 specifies the roman/normal (n) family member for the medium (m) series and the italic (it) member for the bold (b) series. The commands in ⟨info⟩ determine the names of the font metric files and how they should be loaded. In this chapter we shall always use one single font metric file per \DeclareFontShape command, but this is not required [LaTeX3 Project 2000]. The <-> basically states that the font should be scaled to the current type size. The ⟨commands⟩ are executed when a member of the family is loaded. Again, we won't need such commands. ☑

16.9 Creating the Font Definition Files

Having created the font encoding files and the font metric files, we have to implement a font definition (.fd) file for each font family. Figure 16.5 lists the essence of the font definition file for the font family of a seriffed font that uses oldstyle (onum) proportional (pnum) numbers. The last three \DeclareFontShape commands for small caps shapes are not listed.

Note that our naming scheme has made it easy to find the name of the font metric file for a given font shape, font series, and feature combination. For example, let's have a look at the font family

`NexusSerif--fonum--fpnum`. This family has members with and without the small caps feature. For the font metric files we simply use `LY1--NexusSerifOT-⟨series/shape⟩--fonum--fpnum` or `LY1--NexusSerifOT-⟨series/shape⟩--fonum--fpnum--fsmcp`, where ⟨series/shape⟩ is `Regular` for n/m (normal/medium), `Bold` for n/b (normal/bold), `Italic` for it/m (italic/medium), and `BoldItalic` for it/b (italic/bold).

Also note that it is easy to generate the content of the font definition file. As a matter of fact, Owens [2006] has already implemented something like that. Our approach requires a bit more work but is a tad more general.

16.10   Implementing a Font Package

We're almost ready to use our fonts in LaTeX. In this section we shall implement the core of a package that provides commands to access our fonts.

Before writing the package it is useful to see if LaTeX can find the font metric files. We could use `kpsewhich` to do this but we can also do this with the `fonttable` package [Wilson 2011], which provides a command called `\fonttable`. If you pass the name of the font metric file to `\fonttable` the command, it will output the glyphs in the font. Figure 16.6 shows an example that was obtained as follows.

```
\fonttable{LY1--NexusSerifOT-Reg--flnum--fpnum}                LaTeX Usage
```

It's encouraging that the `\fonttable` command works because it means that LaTeX can find the font metric file. The following is a bird's eye view of our package.

```
\ProvidesFile{Nexus.sty}                                       LaTeX Usage
          [2011/07/26 v1.0 Nexus Support]
\makeatletter
\RequirePackage[LY1]{fontenc}
\RequirePackage{pgfopts}
...
\makeatother
```

Loading the `fontenc` package sets the default font encoding to LY1. By loading the `pgfopts` package we can do our option parsing in `pgfkeys` style.

In the remainder of this section we shall implement the core of the remaining commands.

16.10.1   *Parsing the Point Size*

In this section we shall implement a package option that allows the user to define the font's point size (type size). For simplicity we shall implement options for 8–13 pt. Implementing other point size options is straightforward. Throughout we shall use the `pgfkeys`-style option parsing techniques that we studied in Chapter 13.

Each point size should have a properly chosen value for the leading.

Figure 16.6

Sample \fonttable output

	′0	′1	′2	′3	′4	′5	′6	′7	
′00x	ₒ 0	€ 1	2	3	/ 4	˙ 5	″ 6	‚ 7	"0x
′01x	fl 8	9	10	ff 11	fi 12	13	ffi 14	ffl 15	
′02x	1 16	ȷ 17	` 18	´ 19	ˇ 20	˘ 21	¯ 22	˚ 23	"1x
′03x	¸ 24	ß 25	æ 26	œ 27	ø 28	Æ 29	Œ 30	Ø 31	
′04x	32	! 33	" 34	# 35	$ 36	% 37	& 38	' 39	"2x
′05x	(40	) 41	* 42	+ 43	, 44	- 45	. 46	/ 47	
′06x	0 48	1 49	2 50	3 51	4 52	5 53	6 54	7 55	"3x
′07x	8 56	9 57	: 58	; 59	< 60	= 61	> 62	? 63	
′10x	@ 64	A 65	B 66	C 67	D 68	E 69	F 70	G 71	"4x
′11x	H 72	I 73	J 74	K 75	L 76	M 77	N 78	O 79	
′12x	P 80	Q 81	R 82	S 83	T 84	U 85	V 86	W 87	"5x
′13x	X 88	Y 89	Z 90	[91	\ 92	] 93	^ 94	_ 95	
′14x	' 96	a 97	b 98	c 99	d 100	e 101	f 102	g 103	"6x
′15x	h 104	i 105	j 106	k 107	l 108	m 109	n 110	o 111	
′16x	p 112	q 113	r 114	s 115	t 116	u 117	v 118	w 119	"7x
′17x	x 120	y 121	z 122	{ 123	| 124	} 125	~ 126	¨ 127	
′20x	Ł 128	' 129	, 130	ƒ 131	„ 132	… 133	† 134	‡ 135	"8x
′21x	136	‰ 137	Š 138	‹ 139	140	Ž 141	^ 142	- 143	
′22x	ł 144	145	146	" 147	" 148	• 149	– 150	— 151	"9x
′23x	152	TM 153	š 154	› 155	156	ž 157	~ 158	Ÿ 159	
′24x	160	¡ 161	¢ 162	£ 163	¤ 164	¥ 165	¦ 166	§ 167	"Ax
′25x	168	© 169	ª 170	« 171	¬ 172	- 173	® 174	175	
′26x	° 176	± 177	² 178	³ 179	180	µ 181	¶ 182	· 183	"Bx
′27x	184	¹ 185	º 186	» 187	¼ 188	½ 189	¾ 190	¿ 191	
′30x	À 192	Á 193	Â 194	Ã 195	Ä 196	Å 197	198	Ç 199	"Cx
′31x	È 200	É 201	Ê 202	Ë 203	Ì 204	Í 205	Î 206	Ï 207	
′32x	Ð 208	Ñ 209	Ò 210	Ó 211	Ô 212	Õ 213	Ö 214	× 215	"Dx
′33x	216	Ù 217	Ú 218	Û 219	Ü 220	Ý 221	Þ 222	223	
′34x	à 224	á 225	â 226	ã 227	ä 228	å 229	230	ç 231	"Ex
′35x	è 232	é 233	ê 234	ë 235	ì 236	í 237	î 238	ï 239	
′36x	ð 240	ñ 241	ò 242	ó 243	ô 244	õ 245	ö 246	÷ 247	"Fx
′37x	248	ù 249	ú 250	û 251	ü 252	ý 253	þ 254	ÿ 255	
	"8	"9	"A	"B	"C	"D	"E	"F	

Our package provides the allowed point sizes as package options and uses them to determine the corresponding leading. We shall implement each package option as a pgfkeys style that executes two keys: a key that records the point size and a key that records the leading.

We start by implementing the keys that record the point size and the leading. We use the key's .store in to specify the command that records the value of the key.

```
\pgfkeys{/nexus/.cd,
        point size/.store in=\nexus@point@size,
        leading/.store in=\nexus@leading}
```
LATEX Usage

Next we implement a style for the package options. Since there's a lot of repetition we wrap up the style definition in a command.

```
% Define a style called #1pt that sets the
% point size to #1pt and the leading to #2pt.
% The key can also be used as a package option.
\def\nexus@point@size@option#1]#2]{%
    \pgfkeys{/nexus/#1pt/.style={%
            /nexus/.cd,point size=#1,leading=#2}%
}
```
LATEX Usage

The next thing is to determine the values for the leading for the point sizes. A rule of thumb states that the leading should be 1.2 times the value of the point size. Using this rule of thumb usually avoids collisions between characters on adjacent lines.

Intermezzo. Choosing a perfect value for the leading is usually impossible if your text contains inline math material. This is why LATEX—TEX really—has a rescue mechanism that adds some stretch between two lines if $d + h$ exceeds a certain limit, where d is the maximum depth of the boxes on the first line and h is the maximum height of the boxes on the next line. Adding the extra stretch may distrurb the overall page aesthetics. Bazargan, and Radhakrishnan [2007] note that character collisions hardly happen, even if you turn the rescue mechanism off. And even if collisions do happen then they usually can be avoided by rearranging the text. Knuth [1990, pp. 78–80] provides the exact details of TEX's rescue mechanism.

The following defines six options for the allowed point sizes for our package. The option names are 8pt, 9pt, ..., 13pt. For each option we choose a reasonable integral value for the leading. We shall use $\lceil 1.2p \rceil$ pt for the leading, where p is the type size. Defining other point size options is easy. Note that \nexus@combination is expanded before we call \nexus@point@size@option.

```
\@for\nexus@combination:=8]10,9]11,10]12,%
                    11]13,12]14,13]15\do{%
    \expandafter\nexus@point@size@option%
            \nexus@combination]%
}
```
LATEX Usage

We continue by calling the `\ProcessPgfPackageOptions` command, which triggers the parsing of the options that are passed to the package that we're implementing at the moment. The command is implemented in the pgfopts package and is explained in Chapter 13.

```
\ProcessPgfPackageOptions{/nexus}
```
LaTeX Usage

For example, when the user uses our package with the option 11pt then `\ProcessPgfPackageOptions` starts parsing the options of our package. This passes the option 11pt to pgfkeys and this calls the style /nexus/11pt, which sets the point size to 11 pt and the leading to 13 pt.

16.10.2 *Loading the Font*

Having defined the point size and the leading we continue by redefining the commands that store the default names of the roman, sans serif, and typewriter font families.

```
\renewcommand*\rmdefault{NexusSerif--fonum--fpnum}
\renewcommand*\sfdefault{NexusSans--fonum--fpnum}
\renewcommand*\ttdefault{NexusTypewriter--fonum--fpnum}
```
LaTeX Usage

By redefining them we make sure that the proper font is loaded when LaTeX's font selection mechanism selects a roman, a sans serif, or a typewriter shape.

We continue by redefining the remaining defaults. Most of these definitions are standard.

Family, shape, and series:

```
\renewcommand\familydefault{\rmdefault}
\renewcommand\shapedefault{n}
\renewcommand\seriesdefault{m}
```
LaTeX Usage

Medium and bold series:

```
\renewcommand\mddefault{m}
\renewcommand\bfdefault{b}
```
LaTeX Usage

Roman, small caps, italics, and slanted shapes: *Nexus* doesn't have slanted shapes. We could use otftotfm to compute a slanted shape from the roman shape but we shall not do this. Instead we shall use an italic shape when a slanted shape is required.

```
\renewcommand\updefault{n}
\renewcommand\scdefault{sc}
\renewcommand\itdefault{it}
\renewcommand\sldefault{it}
```
LaTeX Usage

Having redefined the defaults, we must make sure that the command `\normalsize` loads the current font at the normal size. It is implemented by setting the current point size and leading with the `\fontsize` command and then calling `\selectfont`.

```
\renewcommand*\normalsize{%                          LaTeX Usage
    \fontsize{\nexus@point@size pt}{\nexus@leading pt}%
    \selectfont%
}
```

We may as well load our \familydefault font family, \shapedefault shape, \seriesdefault series, for the normal point size and leading. Calling to \normalsize should do the trick because its call to \fontsize sets the point size and the leading and its call to \selectfont loads the font.

```
\normalsize                                          LaTeX Usage
```

Redefining \normalsize is important. The main class always defines/redefines the command \normalsize. Therefore you should make sure that you load the main class before you redefine \normalfont. If you use the font package in a user-defined class this means that you should load the main class before you load the font package.

16.10.3 *Changing the Features*

At this stage, we can use the default font family and we can change the font family with the command \fontfamily. However, using the command directly may lead to inconsistencies because it changes two number features at a time. It would be nicer if we could switch a single font feature. In the remainder of this section we shall implement commands that let us do this.

We shall start by implementing a command that defines a command that activates a given font family. This command will be called once for each font family.

```
% #1 is the expanded name of the font family.      LaTeX Usage
\def\nexus@declare@font@family#1\nexus@end{%
    \expandafter\nexus@declare@font@family@%
            \csname#1\endcsname{#1}%
}

% #1 is the command that activates the font family.
% #2 is the expanded name of the font family.
\newcommand*\nexus@declare@font@family@[2]{%
    \newcommand#1{%
        \not@math@alphabet#1\relax%
        \fontfamily{#2}%
        \nexus@renew@defaults#2\nexus@end%
        \selectfont%
    }%
}
```

The idom used by \nexus@declare@font@family@ to change the font is standard. See for example [Lehman 2004, page 43]. When this macro is used, the first argument is the command sequence of the second argument, which is the name of the font family. The macro

`\nexus@renew@defaults` redefines the defaults of the roman, sans serif, and typewriter shapes. The argument of `\nexus@renew@defaults` is always of the form Nexus⟨shape⟩--⟨features⟩.

```
\def\nexus@renew@defaults Nexus#1--#2\nexus@end{%        LaTeX Usage
    \renewcommand*\rmdefault{Nexus#1--#2}%
    \renewcommand*\sfdefault{NexusSans--#2}%
    \renewcommand*\ttdefault{NexusTypewriter--#2}%
}
```

Next we call the macro for our 12 font families. The following is one way to do this. Many font family names are omitted for brevity.

```
\@for\nexus@family:=%                                    LaTeX Usage
    NexusSerif--flnum--fpnum,NexusSerif--flnum--ftnum, ...,
    NexusSans--fonum--fpnum,NexusSans--fonum--ftnum\do{
    \expandafter%
    \nexus@declare@font@family\nexus@family\nexus@end%
}
```

Next let's implement a declaration that keeps the features of the current font family but switches to a sans serif family. Switching to other families may be implemented in a similar way.

```
\newcommand*\SansSwitch[0]{%                             LaTeX Usage
    \nexus@switch@typeface[Sans]%
}
```

In the following implementation of `\nexus@switch@typeface` we use the command `\edef` to expand the family default because `\expandafter` only expands once, which may not be enough.

```
\def\nexus@switch@typeface{%                             LaTeX Usage
    \edef\nexus@next@font{\familydefault}%
    \expandafter\nexus@switch@typeface@\nexus@next@font%
}
```

Note that the command doesn't take any arguments, so calling it as `\nexus@switch@typeface[⟨shape⟩]` effectively inserts the name of the family default before the option `[⟨shape⟩]` and then calls the macro `\nexus@switch@typeface@`: `\nexus@switch@typeface@⟨family default⟩[⟨shape⟩]`. The macro `\nexus@switch@typeface@` is now easy to implement.

```
\def\nexus@switch@typeface@#1--#2[#3]{%                   LaTeX Usage
    \nexus@switch@font{Nexus#3--#2}%
}
```

As the name suggests, the macro `\nexus@switch@font` switches the font. The name of the argument is the name of the new font family.

```
\newcommand*\nexus@switch@font[1]{%                      LaTeX Usage
    \csname#1\endcsname%
}
```

The macro \nexus@switch@font simply turns the name of the font family into one of the command sequences that are defined in the \@for loop earlier on and then executes the command sequence.

We shall now implement a command that turns on the feature onum. Commands that turn on the remaining number features can be implemented in a similar way.

```
\newcommand\ONumSwitch{\nexus@switch@fst[onum]}
```
LaTeX Usage

The command \nexus@switch@fst is implemented using a similar technique as \nexus@switch@typeface.

```
\newcommand\nexus@switch@fst{%
    \edef\nexus@next@font{\familydefault}%
    \expandafter\nexus@switch@fst@\nexus@next@font%
}
```
LaTeX Usage

The option of \nexus@switch@fst@ is the feature that corresponds to the first feature position in the font family. The command substitutes the new feature for the old feature.

```
\def\nexus@switch@fst@#1--f#2--#3[#4]{%
    \nexus@switch@font{#1--f#4--#3}%
}
```
LaTeX Usage

The low-level programming is now done. In the remainder we shall tidy up some loose ends. We start by defining high-level text commands, which are commands that take an argument and typeset the argument in a certain type style. You define such commands with the method \DeclareTextFontCommand. The first argument of this command is the text font command. The second argument is a series of declarations that determine the type style.

```
\DeclareTextFontCommand\SerifStyle{\SerifSwitch}
\DeclareTextFontCommand\SansStyle{\SansSwitch}
...
\DeclareTextFontCommand\TNumStyle{\TNumSwitch}
```
LaTeX Usage

Finally, we redefine the commands \ae, \AE, \o, and \O at the start of the document—it is not clear why this is needed. This is done as follows.

```
\AtBeginDocument{%
    \renewcommand*\ae{\char'032}
    \renewcommand*\AE{\char'035}
    \renewcommand*\o{\char'034}
    \renewcommand*\O{\char'037}
}
```
LaTeX Usage

16.11 Using the Fonts

At this stage it is time for a celebration because we've implemented the core of a package that provides access to the glyphs in an OpenType font by their features. The techniques presented in this chapter may also be used to access other font features and that's how this book was typeset. The following is an example.

```
\noindent%
\texttt{Hello world.}\\          Hello world.
\textsf{Hello world.}\\          Hello world.
\textsc{Hello world.}\\          HELLO WORLD.
\textrm{Hello world.}\\          Hello world.
Hello world.\\                   Hello world.
1234567890.\\\LNumSwitch         1234567890.
1234567890.\\\TNumSwitch         1234567890.
1234567890.\\\ONumSwitch         1234567890.
1234567890.\\\SansSwitch         1234567890.
Pack my box....\\\MixSwitch      Pack my box....
Pack my box....\\\SerifSwitch    Pack my box....
Pack my box....\\\AaltSwitch     Pack my box....
P@ack@@ my@@ box@@....           Pack my box....
```

The last line in the input deserves some explanation because of the @ characters in the input. As you can see from the output, some of the glyphs have swashes and they are accessed with the @ character: ⟨letter⟩@ gives access to the first stylistic alternate of ⟨letter⟩, and ⟨letter⟩@@ gives access to the second stylistic alternate of ⟨letter⟩. This is implemented by defining ⟨letter⟩/@ and ⟨letter⟩@/@ as ligature-forming pairs. Further information may be found in [Toledo 2000] and [Kohler 2011].

References
and
Bibliography

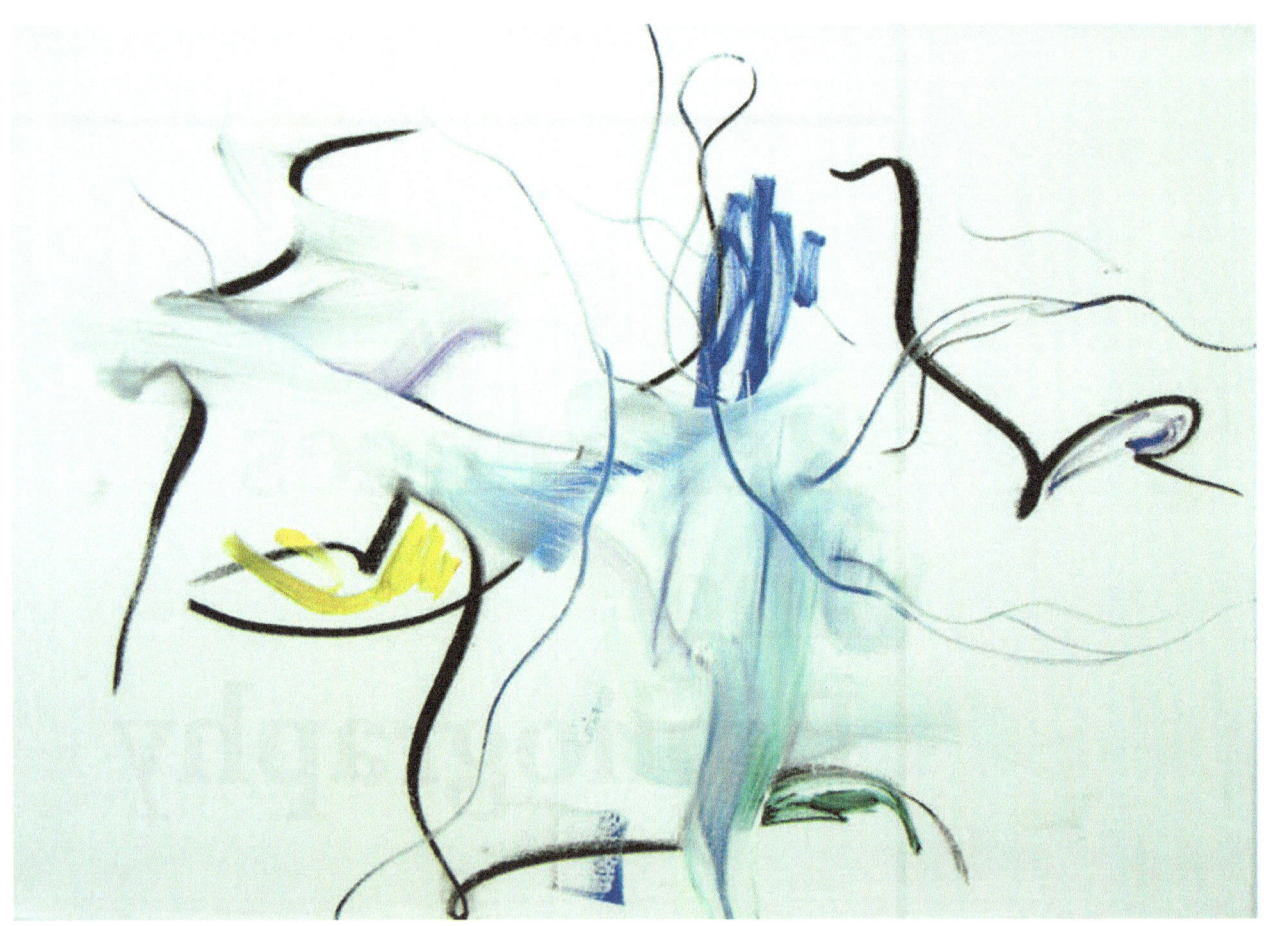

Oil and charcoal on canvas (08/10/10), 64 × 91 cm
Work included courtesy of Billy Foley
© Billy Foley (www.billyfoley.com)

Typographic Jargon

THIS IS A SHORT REFERENCE of typographic jargon. The presentation is not top-down but lists the jargon in alphabetical order. More detailed information may be found in good typography books such as, for example, [Felici 2012].

baseline The baseline is the virtual reference line that the characters are written on. Parts of some of the letters may stick out below the baseline. For example, letters like *g*, *j*, and *p* stick out below the baseline. Figure 16.7 provides an example.

bounding box The bounding box of a character is the virtual rectangle containing the character. Each box has a reference point, a height, a width, and a depth. See Figure 16.8 for an example. The boxes are designed to combine their characters on the baseline by chaining their boxes. This explains why some characters are not contained entirely by their bounding boxes. Figure 16.9 shows how the boxes from Figure 16.8 are combined.

character A character is a member of a typeface.

em The relative length unit em is a synonym for the current type size.

em space An em space is a horizontal space with a length of 1 em.

en An en is a relative length unit that is equal to half an em.

font A computer font is the software that defines a typeface.

font family A font family is a collection of fonts that define a typeface family.

glyph A gyph is a certain form of a character. Different characters usually have different glyphs but the converse is not true. For example, the letter *a* has several different glyphs: a, A, a, A, *a*, *A*, a, A, *a*, *a*, and so on.

italic typeface An italic typeface is a slanted typeface that is usually based on calligraphic handwriting.

kerning Kerning refers to adjusting the space between adjacent glyph pairs that are too close or too far apart. Kerning is needed because the bounding box of a glyph may not always be compatible with the bounding box of any other glyph. When this happens kerning adjusts the position of the glyphs by moving the relative position of their bounding boxes. Figure 16.11 shows how kerning works. In this example kerning reduces the distance of the glyph pairs but kerning may also increase the distance. Kerning may also move glyphs up or down.

Figure 16.7

Baseline and mean line. The solid line is the baseline, which the letters/glyphs are written on. The dashed line is the mean line. The mean line limits the height of the nonascending lowercase letters. Uppercase letters and letters such as *f*, *h*, and *k* stick out above the mean line. Letters such as *g* and *p* stick out below the baseline. The distance between baseline and mean line is the x-height.

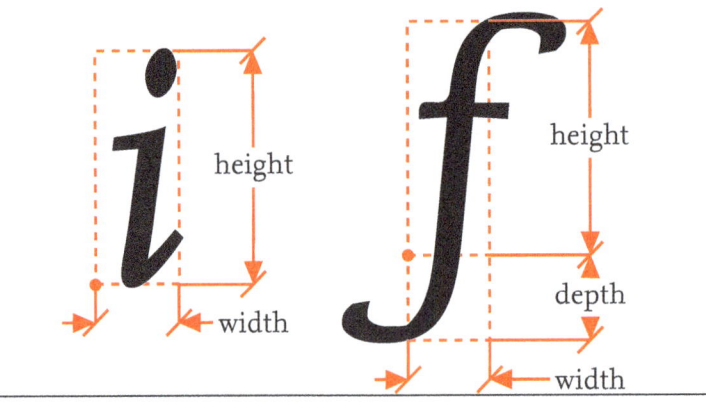

Figure 16.8

Bounding boxes. Each letter is contained in its bounding box, which is drawn with a dashed line. Words are created by sequencing letter boxes (see Figure 16.9). Letters may extend beyond the box outline if this makes it easier to sequence the boxes. The circle on the baseline is the box's reference point. Each box has a height, a width, and a depth. The height (depth) measures the part above (below) the baseline.

Figure 16.9

Word formation. The word 'if' is formed by sequencing the bounding boxes of the letters *i* and *f*. The right side of the bounding box of the *i* touches the left side of the bounding box of the *f* and the reference point of the bounding boxes are on the same horizontal line.

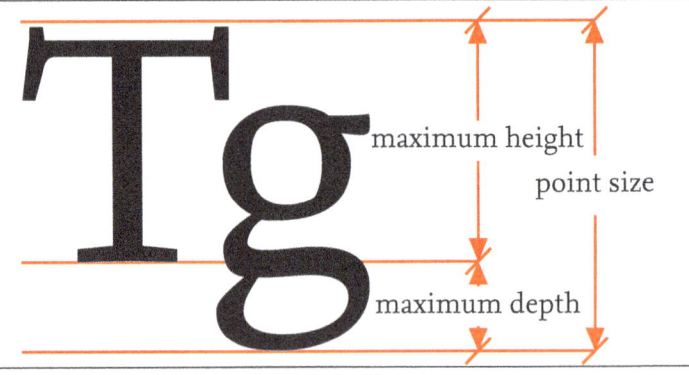

Figure 16.10

Computing the point size. The point size is determined by the sum of the largest possible height and the largest possible depth of the letters in the font at the current size.

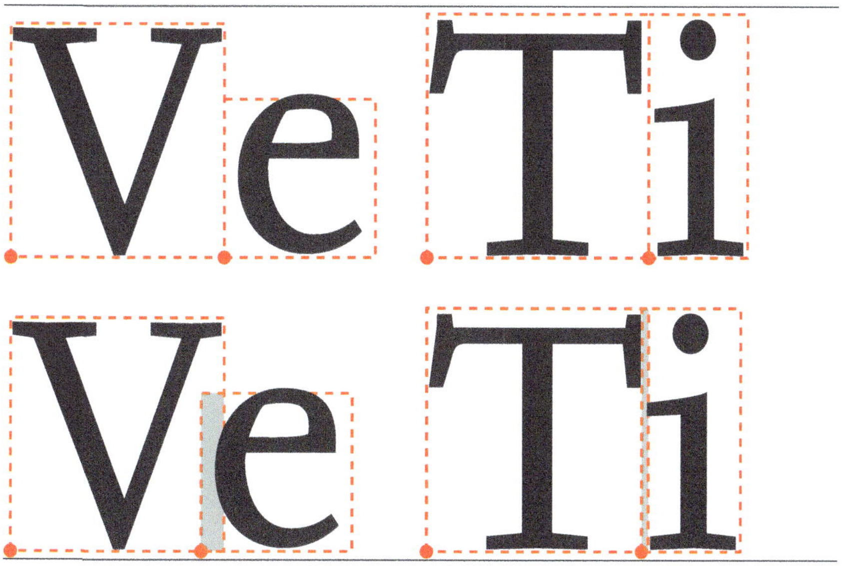

Figure 16.11
Kerning. The distance between the letters in the pairs 'Ve' and 'Ti' on the first line is too large. In the second line this is resolved by adjusting the distances within the pairs. This is called kerning. In this example, kerning reduces the distance between the reference points of the bounding boxes, which makes the bounding boxes overlap. The overlap is shown in grey.

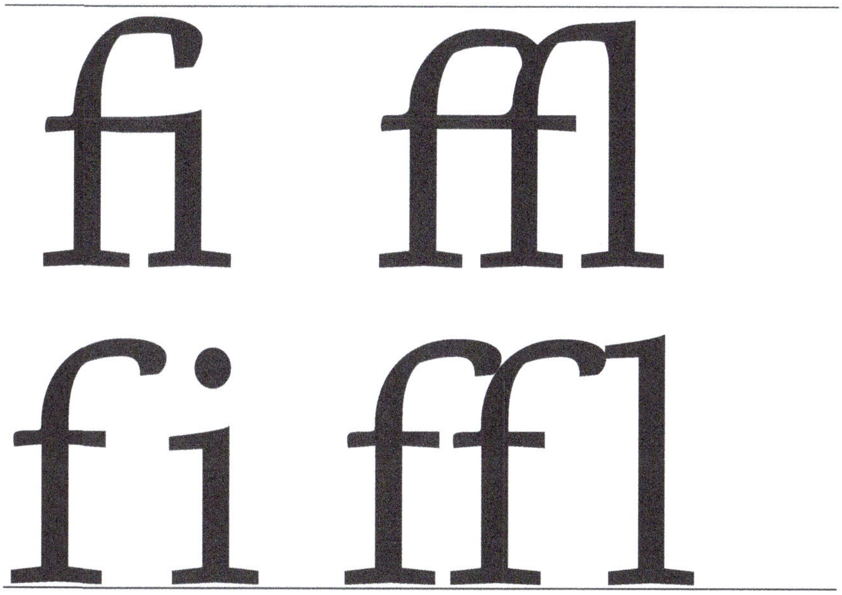

Figure 16.12
Ligatures. This figure shows two example of ligatures. At the top are the ligatures and at the bottom the letters that form the ligatures.

leading The leading or line spacing is the distance between the baselines of two subsequent lines in the text.

letterspacing Letterspacing or tracking refers to the uniform amount of space that is added to the left and right of the characters in a passage of text.

ligature A ligature is a glyph that represents a combination of two or more individual letters, digits, or punctuation marks. Figure 16.12 shows two examples of ligatures.

line spacing See leading.

mean line The mean line is the virtual demarcation line for the height of the lowercase letters without ascenders. See Figure 16.7 for an example.

oblique typeface See slanted typeface.

roman typeface A roman or upright typeface is a typeface whose glyphs

Figure 16.13

Seriffed versus sans serif letters. The letters on the left have serifs, which are drawn in black. The rest of the letters is drawn in grey. The letters on the right lack serifs.

have an upright shape. Usually the main text of papers and books is typeset in a roman typeface.

slanted typeface A slanted or oblique typeface is a typeface whose glyphs are slanted to the right. The resulting glyphs may be obtained by mathematically transforming the roman glyphs. However, some typefaces have a different design.

point A TeX point is an absolute length unit that is equivalent to $1/72.27$ inch. The point unit is denoted pt, so 72.27 pt is equal to 1 inch.

point size The point size or type size is a relative length unit. The point size depends on the font and current size. You get the point size of the font (at the given size) by taking the letters of the font and adding the largest height and the largest depth of the bounding boxes of the letters. Figure 16.10 illustrates the definition.

pt See point.

quad A LaTeX quad is a relative horizontal length unit that is equivalent to 1 em [Knuth 1990, page 314].

sans serif typeface A sans serif typeface is a typeface whose glyphs lack serifs. See Figure 16.13 for an example.

small caps Small caps letters have a similar shape as uppercase letters but they aren't as high so they blend in better with lowercase letters: abcd...xyx ABCD...XYZ.

serif A serif is a small decoration at the end of the strokes of some of the glyphs. See Figure 16.13 for an example.

seriffed typeface A seriffed typeface is a typeface (some of) whose characters have serifs.

thin space A thin space is usually set to ¼ em or about half the word space [Felici 2012]. In LaTeX a thin space is normally ⅙ em [Knuth 1990, page 167].

tracking See letterspacing.

type The word type literally means form and refers to the metal blocks that were used for printing. Nowadays it refers to printable characters.

typeface A typeface is a collection of letters, digits, punctuation symbols, and other characters that share a common design and common features.

typeface family A typeface family is a collection of typefaces that share some features. Many typeface families come with a roman shape, an italic shape, a slanted shape, and so on. Examples of typeface families are *Computer Modern Roman, Computer Modern Sans Serif, Gentium, Linux Libertine Serif/Times New Roman, T$_{\!E}$X Gyre Heros/Helvetica, T$_{\!E}$X Gyre Pagella/Palatino Linotype*, and so on.

type size See point size.

typewriter typeface A typewriter or monospaced typeface is a typeface whose glyphs have equal width.

x-height The x-height is the distance between the mean line and the baseline. By design the distance coincides with the height of the lowercase *x*. See Figure 16.7 for an example.

Bibliography

Abrahams, P. A., K. A. Hargreaves, and K. Berry [2003]. *TeX for the Impatient*. Addison–Wesley. URL: ftp://tug.org/tex/impatient.

Allen, Robert [2001]. *Punctuation*. Oxford University Press. ISBN: 0-19-860439-4.

American Mathematical Society [5th Feb. 2002]. *User's Guide for the amsmath Package*. Version 2.0.

——— [5th Feb. 2002]. *Using the amsthm Package*. Version 2.20.

Arnold, Doug [25th Jan. 2010]. *The LaTeX for Linguists Home Page*. URL: http://www.essex.ac.uk/linguistics/external/clmt/latex4ling/.

Arseneau, Donald [20th Jan. 2010]. *url.sty*. Version 3.3. Edited as a LaTeX document by Robin Fairbairns.

Aslaksen, Helmer [1993]. "Ten TeX Tricks for the Mathematician". In: *TUGboat* 14. A modern version of this paper is available from http://www.math.nus.edu.sg/aslaksen/cs/tug-update.pdf, 135–136.

Bazargan, Kaveh, and C. V. Radhakrishnan [2007]. "Removing vertical stretch—mimicking traditional typesetting with TeX". In: *TUGboat* 28.1, 133–136. URL: http://www.tug.org/TUGboat/tb28-1/tb88bazargan.pdf.

Beccari, Claudio [1997]. "Typesetting Mathematics for Science and Technology According to ISO 31/XI". In: *TUGboat* 18.1, 39–48. URL: http://http://www.tug.org/TUGboat/tb18-1/tb54becc.pdf.

Becker, T., and V. Weispfenning [1993]. *Gröbner Bases* A Computational Approach to Commutative Algebra. Graduate Texts in Mathematics. Springer. ISBN: 3-540-97971-9.

Berry, Karl [1990]. "Filenames for Fonts". In: *TUGboat* 11.4.

Bigwood, Sally, and Melissa Spore [2003]. *Presenting Numbers, Tables, and Charts*. Oxford University Press. ISBN: 0-19-860722-9.

Borowski, Eprahim J., and Jonathan M. Borwein [2005]. *Collins Dictionary of Mathematics*. Collins. ISBN: 0-00-780080-0.

Bovani, Michel [30th Jan. 2005]. *Fourier GUTenberg*.

Breitenbucher, Jon [2005]. "LaTeX at a Liberal Arts College". In: *PracTeX Journal* 3. URL: http://tug.org/pracjourn/2005-3/breitenbucher.

Bringhurst, Robert [2008]. *The Elements of Typographic Style*. Hartley & Marks. ISBN: 0-988179-206-3.

Buchsbaum, Arthur, and Francisco Reinaldo [2007]. "A Tool for Logicians". In: *The PracTEX Journal* 3. URL: http://tug.org/pracjourn/2007-3/buchsbaum.

Burt, John [2005]. "Using poemscol for Critical Editions of Poetry". In: *PracTEX Journal* 3. URL: http://tug.org/pracjourn/2005-3/burt.

Carlisle, David [5th Mar. 1999a]. *The enumerate Package*. Version 3.00.

——— [16th Mar. 1999b]. *The keyval Package*. Version 1.13.

——— [28th May 2001]. *The dcolumn Package*. Version 1.06.

Carlisle, D. P. [14th Nov. 2005]. *Packages in the graphics Bundle*.

Carlisle, D. P., and S. P. Q. Rahtz [16th Feb. 1999]. *The graphicx Package*. Version 1.0f.

Clapham, Christopher, and James Nicholson [2005]. *Oxford Concise Dictionary of Mathematics*. Third Edition. Oxford University Press. ISBN: 978-0-19-860742-7.

Dashboard, Humanities Division [8th May 2008]. *LATEX—Humanities Computing*. URL: https://coral.uchicago.edu:8443/display/humcomp/LaTeX.

Dearborn, Elizabeth [2006]. "TEX and Medicine". In: *PracTEX Journal* 4. URL: http://tug.org/pracjourn/2006-4/dearborn.

Eijkhout, V. [2007]. *TEX by Topic, A TEXnician's Reference*. Addison–Wesley. ISBN: 0-201-56882-9. URL: http://www.eijkhout.net/tbt/.

Felici, James W. [2012]. *The Complete Manual of Typography* A Guide to Setting Complete Type. Second Edition. Adobe Press. ISBN: 978-0-321-77326-5.

Fenn, Jürgen [2006]. "Managing Citations and Your Bibliography with BibTEX". In: *PracTEX Journal*. URL: http://www.tug.org/pracjourn/2006-4/fenn/.

Feuersänger, Christian [5th Aug. 2010a]. *Manual for Package* PGFPLOTS. Version 1.4.1.

——— [5th Aug. 2010b]. *Manual for Package* PGFPLOTSTABLE. Version 1.4.1.

Fine, Jonathan [1992]. "Some Basic Control Macros for TEX". In: *TUGboat* 13.1.

Fiorio, Christophe [14th Dec. 2004]. algorithm2e.sty—*package for algorithms*. Version 3.3.

Flynn, Peter [2007]. "Rolling your own Document Class: Using LATEX to keep away from the Dark Side". In: *TUGboat* 28.1. URL: http://www.tug.org/TUGboat/Articles/tb28-1/tb88flynn.pdf.

Garcia, Aracele, and Arthur Buchsbaum [2010]. "About LATEX tools that students of Logic should know". In: *PracTEX Journal* 1. In Portugese. URL: http://tug.org/pracjourn/2010-1/garcia.

Giacomelli, Roberto [12th July 2009]. calctab *Package*. Version 0.6.1.

Goldberg, Jeffrey P., and H.-Martin Münch [1st Feb. 2011]. *The* lastpage *Package*. Version 1.2g.

Goossens, M., S. Rahtz, and F. Mittelbach [1997]. *The LATEX Graphics Companion: Illustrating documents with TEX and PostScript*. Addison–Wesley. ISBN: 0-201-85469-4.

Graham, R. L., Donald E. Knuth, and O. Patashnik [1989]. *Concrete Mathematics: A Foundation for Computer Science*. Addison-Wesley. ISBN: 0-201-14236-8.

Hedrick, Charles [15th Jan. 2003]. *Guidelines for Typography in* NBCS. URL: http://www.nbcs.rutgers.edu/~hedrick/typography/typography.janson-syntax.107514.pdf.

Heinz, Carsten, and Brooks Moses [22nd Feb. 2007]. *The* Listings *Package*. Version 1.4.

Heldoorn, Marcel [2nd Dec. 2007]. *The* SIunits—*package* Consistent application of SI Units. Version 1.36.

Høgholm, Morten [28th July 2008]. *The* breqn *Package*. Version 0.98.

———— [17th Jan. 2011]. *The* xfrac *Package*. Version 0.3b.

Høgholm, Morten et al. [12th Feb. 2011]. *The* mattools *Package*. Version 1.10.

Kern, Uwe [21st Jan. 2007]. *Extending* LaTeX's *color facilities:* xcolor. Version 2.11.

Knuth, Donald E. [1990]. *The* TeXbook. Addison–Wesley. ISBN: 0-201-13447-0. URL: http://www.ctan.org/tex-archive/systems/knuth/tex/.

Kohler, Eddie [2011]. otftotfm *Manual*. Version 2.90. URL: http://www.lcdf.org/type/otftotfm.1.html.

Krab Thorub, Kresten, Frank Jensen, and Chris Rowley [22nd Aug. 2007]. *The* calc *Package,* Infix Notation Arithmetic in LaTeX.

Lamport, Leslie [1994]. LaTeX: *A Document Preparation System*. Addison–Wesley. ISBN: 0-021-52983-1.

LaTeX3 Project [12th Mar. 1999]. LaTeX2ε *for Class and Package Writers*. URL: http://www.latex-project.org/guides/clsguide.pdf.

LaTeX3 Project [9th Feb. 2000]. LaTeX2ε *Font Selection*. URL: http://www.latex-project.org/guides/fntguide.pdf.

Lehman, Philipp. *The* biblatex *Package*. Version 1.5.

———— [2004]. *The Font Installation Guide* Using Postscript Fonts to their Full Potential in LaTeX. Version 2.14.

Longborough, Brent [28th Aug. 2011]. tagging.sty. Version 1.0.

Midtiby, Henrik Skov [21st Apr. 2011]. *The* todonotes *Package*.

Miede, André [24th Jan. 2010]. *The Classic Thesis Style*.

Mittelbach, Frank, Robin Fairbairns, and Werner Lemberg [6th Jan. 2006]. LaTeX *Font Encodings*.

Mühlich, Matthias [23rd Feb. 2006]. *The* CoverPage *Package*. Version 1.01.

Oberdiek, Heiko [23rd Dec. 2010]. *The* keyvalue *Package*. Version 3.10.

Ochsenmeier, Erwin [2011]. FourSenses.net. At the moment of writing, this site is being moved from http://theotex.blogspot.com/. URL: http://http://en.foursenses.net/.

Oetiker, Thomas [8th Sept. 2010]. *An Acronym Environment for* LaTeX2ε. Version 1.36.

Oetiker, Tobias et al. [2007]. *The Not so Short Introduction to* LaTeX2ε. URL: http://tobi.oetiker.ch/lshort/.

Owens, John D. [2006]. "The Installation and use of OpenType Fonts in LaTeX". In: *TUGboat* 27.2, 112–119.

Pakin, Scot [2005]. *The Comprehensive LATEX Symbol List.* URL: `http://tug.ctan.org/info/symbols/comprehensive/symbols-letter.pdf`.

Patieri, Lórenzo [2010]. *Customizing classicthesis with the arsclassica Package.*

Peyton Jones, S., and J. Hughes [1999]. `Haskell 98:` *A Non-strict, Purely Functional Language.* `http://www.haskell.org/onlinereport/`.

Peyton Jones, Simon L., John Huges, and John Launchberry [1993]. "How to Give a Good Research Talk". In: *ACM SIGPLAN Notices* 28.11, 9–12. URL: `http://research.microsoft.com/en-us/um/people/simonpj/papers/giving-a-talk/giving-a-talk.htm`.

Rahts, Sebastian [1993]. "Essential NFSS2, Version 2". In: *TUGboat* 14.2, 132–137.

Reckdahl, Keith [2006]. *Using Imported Graphics in LATEX and pdfLATEX.* URL: `ftp://ftp.tex.ac.uk/tex-archive/info/epslatex.pdf`.

Reichert, Axel [4th Aug. 1998]. *units.sty—nicefrac.sty.* Version 0.9b.

Robertson, Will [26th Feb. 2011]. *The fontspec Package.* Version 2.1f.

Scharrer, Martin [2011-07-23]. *List of Internal LATEX2ε Macros useful to Package Authors.* Version 0.4.

Schlicht, R. [10th Jan. 2010]. *The microtype Class.* Version 2.4.

Schneider, Tom [27th Apr. 2011]. *LATEX Style and BIBTEX Bibliography Formats for Biologists: TEX and LATEX Resources.* URL: `http://www-lmmb.ncifcrf.gov/~toms/latex.html`.

Senthil, Kumar M. [2007]. "LATEX Tools for Life Scientists (BioTEXniques?)" In: *PracTEX Journal* 4. URL: `http://tug.org/pracjourn/2007-4/senthil`.

Smith, Peter. *LATEX for Logicians.* URL: `http://www.logicmatters.net/latex-for-logicians/`.

Strunk, W., and E. B. White [2000]. *The Elements of Style.* Fourth Edition. Macmillan Publishing. ISBN: 0-205-30902-x.

Talbot, Nicola [15th Nov. 2009]. *datatool v 1.01: Database and data manipulation.* Version 2.03.

Tanksly, Charlie. *LATEX: A Guide for Philosophers.* URL: `http://www.charlietanksley.net/latex/`.

——— *PhilTEX Forums* A Place for Philosophers to Learn about LATEX. URL: `http://www.charlietanksley.net/philtex/forum/`.

Tantau, Till [25th Oct. 2010]. *TikZ & PGF.* Version 2.00-cvs.

Tantau, Till, Joseph Wright, and Vedran Miletić [12th July 2010]. *The BEAMER Class.* Version 3.10.

Taraborelli, Dario [2010]. *The Beauty of LATEX.* URL: `http://nitens.org/taraborelli/latex`.

Tellechea, Christian [8th Apr. 2011]. *spreadtab.* Version 0.3c.

Thomson, Paul A. [2008a]. "Clinical Trials Management on the Internet—I. Using LATEX and SAS to Produce Customized Forms". In: *PracTEX Journal* 3.

——— [2008b]. "Clinical Trials Management on the Internet—II. Using LATEX, PostScript, and SAS to Produce Barcode Label sheets". In: *PracTEX Journal* 3. URL: `http://tug.org/pracjourn/2008-3/thompson2`.

Toledo, Sivan [2000]. "Exploiting Rich Fonts". In: *TUGboat* 21.2.

Trask, R. L. [1997]. *Penguin Guide to Punctuation*. Penguin Books. ISBN: 0-140-51366-3.

Tufte, Edward R. [2001]. *The Visual Display of Quantitative Information*. Second Edition. Graphics Press LLC. ISBN: 978-0-9613921-4-7.

Turabian, Kate L. [2007]. *A Manual for Writers of Research Papers, Theses, and Dissertations*. Seventh. University of Chicago Press. ISBN: 978-0-226-82337-9.

Turner, Ken [1st Nov. 2010]. *BibTeX Style Examples*. URL: http://www.cs.stir.ac.uk/~kjt/software/latex/showbst.html.

Unger, Gerard [2007]. *While You're Reading*. Mark Batty Publisher. ISBN: 978-0-9762245-1-8.

Van Oostrum, Piet [2nd Mar. 2004]. *Page Layout in LaTeX*.

Veytsman, Boris, and Leila Akhmadeeva [2006]. "Drawing Medical Pedigree Trees with TeX and PSTricks". In: *PracTeX Journal* 4. URL: http://tug.org/pracjourn/2006-4/veytsman.

Voß, Herbert [2010]. *Math Mode*. URL: ftp://cam.ctan.org/tex-archive/info/math/voss/mathmode/Mathmode.pdf.

Wilson, Peter [13th Feb. 2011]. *The fonttable Package*. Version 1.6b. Maintained by Will Robertson.

Wilson, Peter, and Lars Madsen [6th Mar. 2011]. *The Memoir Class*.

Wright, Joseph [1st May 2010]. *pgfopts—LaTeX Package Options with pgfkeys*. Version 2.0.

———— [15th June 2011]. *siunitx—A Comprehensive (SI) units package*. Version 2.2i.

Zeidler, Eberhard, ed. [1996]. *Oxford User's Guide to Mathematics*. Typeset in LaTeX by Bruce Hunt. Oxford University Press. ISBN: 0-19-850763-1.

Acronyms and Abbreviations

AMS	American Mathematical Society
API	Application Programming Interface
APL	A Programming Language
CTAN	Comprehensive TeX Archive Network
CD	Compact Disk
FAQ	Frequently Asked Question
GUI	Graphical User Interface
IDE	Integrated Development Environment
ISBN	International Standard Book Number
SI	Système International d'Unités/International System of Units
OS	Operating System
TUG	TeX Users Group
URL	Uniform Resource Locator
WYSIWYG	What You See is What You Get

Indexes

LᴬTᴇX and TᴇX Commands

Environments

Classes

Packages

Languages and External Commands

Colophon

THIS BOOK was created with `pdflatex` in a standard TeX Live installation. The cover, the spine, and pages i–iv were produced by the publisher. The main text was typeset with the `book` class, using *FF Nexus* at $11/13 \times 27$ as the font family, and scaling its typewriter fonts to 83%.

The page, figure, and table layout were implemented with a user-defined package. The same holds for the `itemize`, `enumerate`, and `description` environments.

I had two main concerns when designing the page layout. First, I wanted figures and tables that could run into the margins. With program listings this is almost always needed; also this would let me typeset input and output side by side. My second concern was that I wanted the figure and table captions to the side. That way, long explanations would not be so disruptive.

The artwork at the back of the part titlepages is included courtesy of Billy Foley, a Cork-based artist and member of the Cork Artists Collective. The landscape on page 2 is included courtesy of the University College Cork Art Collection. More of Billy Foley's work may be found on `www.billyfoley.com`.

The `amsmath` and `amssymb` packages were used to help typeset the mathematics. The `mathastext` package was used with the option `italic` to make sure that the numbers and letters in mathematical expressions were typeset in *Nexus*. The result is not always perfect but overall it looks pleasing. The bibliography was typeset with the `biblatex` package. The `microtype` package was used with the options `tracking=smallcaps`, `expansion=true`, and `protrusion=true`.

The manufacturer's authorised representative in the EU is Springer
Nature Customer Service Centre GmbH, Europaplatz 3, 69115 Heidelberg,
Germany. If you have any concerns regarding our products, please
contact ProductSafety@springernature.com

Printed and bound by CPI Group (UK) Ltd, Croydon, CR0 4YY

28/04/2026

02098462-0005